P9-DFX-392

THE LETTS GUIDE TO COLLECTING

F A N S

THE LETTS GUIDE TO COLLECTING

F A N S

SUSAN MAYOR

THE LETTS GUIDE TO COLLECTING FANS

Published in 1991 by Charles Letts
London, England

Copyright © 1990 Studio Editions Ltd.
Text © 1990 Susan Mayor

All rights reserved. No part of this publication may be reproduced, stored in a retrieval system, or transmitted in any form or by any
means, electronic, mechanical, photocopying, recording or otherwise, without the prior permission of the copyright holder

This book was edited and designed by Anness Law Limited
4a The Old Forge, 7 Caledonian Road, London N1 9DX

ISBN 1-85238-128-0

Printed and bound in Czechoslovakia

Plates on pages 10 and 19 are by courtesy of the Rt. Hon the Earl of Harewood, K.B.F., Harewood House, Yorkshire. Plates on pages
44, 50, 52, 53, 54, 58 (2), 59, 62, 67, 68, 82 and 87 are by courtesy of the Hon. C. A. Lennox-Boyd. The plate on page 35 is by courtesy
of a private collector. All other photographs courtesy of Christie's South Kensington, London.

Publisher's Note: The exchange rate at which Pounds are converted to US Dollars throughout this book is £1.00 = $1.65.

CONTENTS

FOREWORD

Fans as ceremonial, practical and above all

decorative items have been recorded from

Egyptian times to the 18th century, the great

age of the fan, and to the present day. This

book is a collector's guide to these fascinating

and exquisite accessories.

I t is now ten years since my previous book, *Collecting Fans*, was published. Since that time thousands of fans have been sold at Christie's and a number of important exhibitions have been held. I felt it was time that collectors were brought up to date with new discoveries and unpublished research.

Ten years ago I noted collecting fans was an agreeable disease. That disease threatens to become an epidemic. At Christie's South Kensington we now hold four fan sales a year, each with some three hundred lots. Now, apart from the Fan Circle International, there are three other fan clubs: F.A.N.A., the Fan Association of North America; Associazione Culturale 'Il Venteglio', Bologna, and Cercle de l'Eventail, Paris. Each sponsors conferences, publications and exhibitions.

This book has been designed as an introduction for would-be collectors and incipient experts. It is not a documented study and therefore the bibliography has been made as comprehensive as possible.

In this new edition, the layout has been re-designed, the selection of illustrations has been totally revised in order to incorporate colour photographs and fresh discoveries, and the text has been re-arranged. It now includes material which has come to light during the last decade, and the new illustrations are captioned with prices realized in recent sales. The bibliography has been updated and a glossary, list of exhibitions and an index included. The list of museums has also been extended.

So many people have helped me—either through discussing their collections or by bringing their fans to Christie's South Kensington—that it is impossible to thank them all. However, I must single out Mrs. Madeleine Ginsburg, formerly of the Victoria and Albert Museum; Mrs. Helene Alexander of the Fan Museum, Greenwich; Mrs. Betty Hodgkinson M.B.E., President of the Fan Circle International; Dr. Alice Zriebiec of The Metropolitan Museum, New York; Miss Ann Coleman, of Brooklyn Museum, New York; Dr. van Eeghen, formerly of the Dutch Royal Archives; and The Hon. Christopher Lennox-Boyd who, as always, has been a mine of recondite information.

Left: A beautiful group of fans. From top left, shown closed, a fan with ivory sticks, one guardstick set with a pomander, c.1860; a chromolithographic advertising fan including advertisements for Agence Cook and a view of the Grand Hotel de la Gare du quai d'Orsay, published by J Ganne, 1907; a Canton cockade fan painted with figures on terraces, their faces of ivory, their clothes of silk, with varigated lacquer sticks, mid-19th century; a fan, the leaf painted with an elegant woman fishing, with putti, signed M Rodigue, c.1885; a fine fan, the leaf painted with two ladies and putti in a landscape, signed F Houghton, the smokey mother-of-pearl sticks pierced and gilt, c.1880; a fan, the black lace mount worked with a peacock in a formal garden, the tortoiseshell sticks pierced and piqué with gold, c.1870.

A HISTORY OF FAN COLLECTING

Fan collecting has gone in waves, peaking in

periods when the accumulation of fans has

become almost a craze.

Fan collectors are a diverse breed. They have ranged from Queen Mary, Consort of King George V—who accumulated a great number of fans during a lifetime of Royal Progresses—to a late nineteenth-century French lawyer called Lucien Duchet, who crammed his chambers with ladies' fans dating back to the tumultuous days of the French Revolution.

There is something about a fan that appeals to a collector: it fits snugly in the hand, its colour and pattern are an endless delight, and its provenance conjures up images of elegant courts and Bals masqués.

There appear to have been two main periods of fan collecting: the 1860s to 1910s in England, and probably also Germany and other European countries; and the 1920s and 1930s in America and England.

The two celebrated collectors of the first period were Mr. Robert Walker and Lady Charlotte Schreiber. Their superb collections were both catalogued, Walker's for his exhibition at the Fine Art Society and sale at Sotheby's in 1882, and Lady Charlotte's by Lionel Cust when given to the British Museum in 1891. Mr. Walker had, in turn,

bought some of his collection from Sir Augustus Wollaston Franks of the British Museum.

Another fan collection from this period was that sold at C.S.K. on 29 July 1976, the property of a nobleman. This consisted of 59 superb fans, and sparked off a new wave of high prices: lot 59 in the sale—a French fan of about 1760 – realized £950 ($1,500). This unusual fan had a leaf entirely covered with large mother-of-pearl panels painted with gods, mortals and putti; the ivory sticks were carved and pierced but not gilt. There were some rare eighteenth-century lace fans and six eighteenth-century cabriolet fans, considered very rare. There were also interesting chinoiserie and Chinese fans and an amusing English fan of about 1755 with figures running in a maze. This collection was certainly eclectic.

Collectors such as Lady Charlotte Schreiber specialized more narrowly. Her collection was, of course, far larger than most others of the period, but she was certainly more interested in the history and subject matter of each fan than in its beauty, and most of her fans were printed. Much of her collection consisted of unmounted leaves. Lady Charlotte (1812–95), who married first in 1833 Josiah John Guest, the founder of what was to become Guest Keen & Nettlefold, and then in 1855 her children's tutor Charles Schreiber, was an avid collector. She kept diaries of her collecting trips or 'chasses', as she called them. Madeleine Ginsburg gave a charming talk about Lady Charlotte's life to members of the Fan Circle at the British Museum in 1979.

Another splendid collection of this generation was that formed by Sir Matthew Digby Wyatt, which forms the basis of the Victoria and Albert Museum's collection.

Above: North European, late 17th century. The sticks are made from carved, pierced tortoiseshell. The leaf is black, painted with a scene of putti organizing the ablutions of a prince. The reverse is painted with bunches of flowers.

Right: A group of fans offered at Christie's South Kensington in November 1989, including: centre, a fan painted with Vulcan's Forge, the verso with chinoiserie, *the ivory sticks carved and pierced with Diana and other figures, c.1730, 11ins (28cms); top right, a fan with ivory sticks, its leaf painted with the Temple* *of Sybil at Rome, Italian, 10ins (25cms), late 18th century; bottom centre, a fan with a leaf of Brussels lace and tortoiseshell sticks carved and set with enamel plaques, French, 14ins (36cms), c.1890, £385 ($635). More fans from this group are identified on page 16.*

Above: English, c.1745. A fan, the leaf painted with Hymen at the Altar of Love, the mother-of-pearl sticks silvered. Formerly from the collection of Her Late Majesty Queen Mary, now in the collection of the Rt Hon the Earl of Harewood.

Above: French, 1783. A ballooning fan with ivory sticks. The leaf is of silk and is decorated with a scene of the balloon ascent of Mm. Robert and Charles, 1st December 1783.

Besides being an authority on the decorative arts of the Middle Ages, Wyatt was an expert in industrial art; in association with Prince Albert, Owen Jones and Henry Cole, he was one of the leading spirits behind the 1851 Exhibition, and a founding father of the Victoria and Albert Museum.

Another collection still in the collector's family was made by the Dowager Marchioness of Bristol. This included a superb fan painted with Hector and Andromache with jewelled guards and a watch set in the pivot, illustrated in Woolliscroft Rhead. A portion of another collection, Lucien Duchet's revolutionary fans collected in the 1880s and 1890s, was sold by Christie's in 1981.

It is often difficult to differentiate between noble collectors like Lady Bristol who searched out new items and those who were just interested in their existing collection of inherited fans, but among the lenders to the Karlsruhe Exhibition in 1891 and the Madrid Exhibition of 1920, possibly Dr. Marc Rosenberg (or rather his brother, the artist and collector Gabriel Jacob Rosenberg) was one of the few who had built up his collection from the start.

The post-war generation is not so well known. The best-known collection was formed by Leonard Messel and was sold to the Fitzwilliam Museum in Cambridge in 1985. He was very fond of Oriental fans but his collection also included many important eighteenth-century European fans.

The collection formed by Mrs. Baldwin of Milwaukee was sold by her daughter Mrs. Pabst at C.S.K. on 4 May 1978; she bought well-painted fans and her collection was strong in commemorative fans. That formed by Mrs. H. Bompas was sold at Christie's on 14 July 1969; this collection was very strong in English eighteenth-century printed fans and included several not in Lady Charlotte Schreiber's collection.

Another collection sold at auction was that made or inherited by the Duquesa de Marchena, sold by C.S.K. in October 1977.

The big modern collectors are nearly all members of the Fan Circle International, which was formed in London in 1975. There are of course many other fan collectors but their collections are usually on a smaller scale.

In the 1880s fan prices were high, possibly higher in real terms than they are now, but by the 1920s they had fallen in price and it is only in the last fifteen years that they have returned to the prices they were fetching at Mr. Robert Walker's sale in 1882. Provenance is important too; the cachet of royal or theatrical ownership always adds appreciably to the price.

WHAT SHOULD A NEW COLLECTOR BUY?

I would suggest browsing around a few auction sales before taking the plunge. The choice is endless—C.S.K. sells over 2,000 fans each year. As it would be very expensive to form a comprehensive collection of fans of all types and all periods, I would suggest that the collector specializes on some theme. But what theme? Collecting ballooning fans would be as frustrating as it is expensive, as general collectors of ballooning ephemera enter this field. It is also inadvisable to enter a field of collecting in which there are already a number of collectors: this often forces the price up without the new collector getting very many fans. The collecting of early printed fans is such a field. Not enough examples of each fan turn up to satisfy the existing collectors. The collector should also avoid specializing in a very rare field like mica fans: only two have come on the market in twenty years, and only a handful of others are known.

If the collector wishes to have a small but expensive collection he could possibly collect eighteenth-century cabriolet, eighteenth-century lace, eighteenth-century *trompe l'œil* or early commemorative fans. If the aim is to have a large attractive collection, pretty eighteenth-century painted fans of lesser importance, pretty late nineteenth-century fans or mid-nineteenth-century lithographic fans should be considered.

THE ORIGINS OF
THE FAN

No-one really knows when the fan came into

being, but since the earliest days when it was

used for wafting air for coolness, it has

developed into one of the most intricate and

decorative art forms.

Fans are as old as hot weather. It is impossible to pinpoint where and when the fan originated. In hot climates it must always have been invaluable—in creating a breeze and keeping flies away—after all, the earliest known man comes from near the Equator. Later it became a work of art and was used in religious and royal ceremonies. Later still, it became the decorative accessory that we know today.

In its earliest form the fan was probably a small hand-sized screen—better and more effective than just fanning oneself with one's hands or a leaf. Handles were then added and these screens were adapted and decorated in numerous ways, using local materials and meeting demands and fashions as they arose.

The earliest fans are depicted in contemporary bas-reliefs, sculpture and painting. In Egypt fan-bearing courtiers carried semicircular feather fans on long poles. Some surviving examples have been found in tombs. That of Tutankhamen (1350 B.C.) contained one chased in gold with scenes of the young king hunting ostriches: their feathers would be used to decorate fans.

Early fans were also used for winnowing (separating the chaff from the grain), fly-swatting and starting fires. Surviving examples of early fans are, of course, extremely rare and their history will have to be filled in gradually from contemporary painting, sculpture and literature.

Probably the earliest Chinese specimens of fans are two woven bamboo side-mounted fans (second century B.C.),

Left: French, 1682. The fan leaf has been dismounted and enlarged. It is painted with the scene of the twentieth birthday of le Grand Dauphin in 1681, including the royal family and the court.

Above: A selection of fans sold at Christie's South Kensington in December 1986, including: an ivory brisé fan painted with a lady fishing, the reserves pierced and gilt, and painted with birds, roses, and honeysuckle, 9ins (23cms), £880 ($1452); a fan with its leaf painted with a view of Bath including figures and sedan chairs in the foreground, with ivory sticks, English, 11½ins (29cms), c.1750, £2200 ($3630); and, shown closed, a fan with tortoiseshell sticks piqué with gold, its leaf painted with Pompeian frescoes of Youth and Age, Italian, 11½ins (29cms), c.1760, £770 ($1270).

excavated from the Mawangdui tomb near Changsha in Hunan province, but fans certainly existed in China long before that date.

The first Chinese fans—hand-screens—were probably of feathers, which would not have survived. At one stage peacock feathers were employed, then, to economize, silk was used and sometimes silk tapestry (*kesi*). They were sometimes curved at the tip to create a better breeze. Both sexes carried fans in China and detailed regulations accorded different types of fan to each rank of person. The fan was used in ceremonies and could also be used to shield one's face when passing dignitaries of equal rank—thus averting the necessity of endless greeting rituals.

Gradually, more and more is being discovered about the origin of the folding fan. It is now thought that the brisé fan originated in Japan and was taken to China by a missionary (or traveller) in the ninth century. The folding fan was then adopted with alacrity, as it was obviously more convenient. When not in use it could be folded away in an embroidered silk fan case which, in turn, could be stored in the cuff, near the neck or in the boot, until next required.

The Western world's earliest surviving fan is preserved in the basilica of St. John the Baptist at Monza, twelve miles from Milan. There, in a domed octagonal treasury, can be seen the *flabellum*—or ceremonial fan—presented to the basilica by Theodolinda, the sixth-century Queen of the Lombards. This unique survival is made of purple vellum and decorated with gold and silver ornaments, like a similar later fan at Florence, known as the Tournus Flabellum. Queen Theodolinda's fan still retains its wooden box and silver-mounted handle. Strangely, it was not widely known until 1857, when its existence was publicized by the Victorian architect William Burges. There is a third known flabellum—that of S. Sabino, a sixth-century bishop—that is still in the treasury of his cathedral in the south of Italy.

DATES AND STYLES

Fans come in an infinite variety of styles and

designs that reflect centuries of changing tastes

in art and fashion; they can be painted, printed,

plain, decorative, European, Oriental,

novelty, paper, feather, large and small.

As fans are essentially decorative novelties, exquisite luxuries, they can be made from every conceivable material providing it is strong. One thing one must never forget: for such small and delicate items, fans have to be extremely tough and hard-wearing. Thousands have stood the test of time: they are not nearly as brittle as they look. During most periods fans were a fashionable accessory and so tended to be made from materials in fashion and available at the time; at some periods so many different materials were used that it is hard to catalogue or even list them all.

Many of the early fans—Egyptian, Chinese, and late sixteenth-century European fans—were hand-screens trimmed with feathers, mainly ostrich, sometimes dyed (hand-screens are pieces of a rigid material of various shapes on the end of a stick). The handles, however, were more varied. Queen Elizabeth I possessed fans with gold handles set with precious stones. They are listed in her wardrobe list in the Public Records Office. Sir Roy Strong and Miss Julia Trevelyan Oman mention five of the most sumptuous from the 1600 list in their book, *Elizabeth R*, including:

Item, one fanne of feathers of divers colours, the handle of golde with a bear and a ragged staffe on both sides, and a lookinge glass on one side. [Presumably a gift from the Earl of Warwick, as the bear and rugged staff was his badge, or his brother Robert Dudley, Earl of Leicester, her favourite.]

Item, one handle of golde enameled, set with small rubies and emeralds lackinge nine stones, with a ship under sail on one side.

Item, one fanne of white feathers, with a handle of golde having two snakes wyndinge about it, garnished with a

Above: *The leaf of the fan is decorated with a scene of elegant shepherds and shepherdesses singing, the ivory sticks* piqué *with silver, carved with figures, and* clouté *with mother-of-pearl, some of which is missing. 11ins (28cms), c.1720, sold at Christie's South Kensington in December 1988 for £825 ($1361).*

ball of diamonds in the ende, and a crowne on each side within a paire of wings garnished with diamonds, lackinge six diamondes.

When Elizabeth I died she left a collection of 27 fans. As she was fond of them, she often carried them in portraits and considered them a suitable present for a Queen. The same type of fan reappeared in the reign of Queen Victoria.

At the French court, folding fans of pierced kid cut to resemble Italian *reticella* lace were fashionable. These had ivory sticks. Very few are known to have survived, but two were discussed by Germain Bapst in 1881. Both were threaded with plaques of mica, so that they would have shimmered when moved. The more elaborate one, from the Revoil collection, measured 26 cm and opened to 45°, with ten ivory sticks. The guardsticks were each trimmed with six silk pompons, similar to those on Japanese fans. The leaf was of vellum *découpé* and slotted into the sticks, alternating with mica panels painted with coloured scenes of Diana and Actaeon in contemporary dress (inspired by the chimneypiece by Hugues Lallemend in the Hôtel de Cluny), and with hearts pierced by an arrow and a flaming torch. Bapst suggests that this fan dated from about 1580. The second fan, a little later in date, from the Sauvagest Collection, was larger (32 cm) and composed of a double leaf of *découpé* vellum with mica and silk inserted between the leaves. At the time (1881), Bapst knew of other examples, including those in the collections of Mme. Jubinal and M. Dupont Auberville. Woolliscroft Rhead, however, suggests that the Jubinal fan was seventeenth-century.

A very similar fan from the collection of the late Miss Esther Oldham is illustrated in Nancy Armstrong's *Collector's History of Fans*. It is probably the one Miss Oldham since gave to the Boston Museum.

From Pierre de l'Estoile's *l'Isle des Hermaphrodites* (1588), Bapst quotes a passage describing Henri III using a similar type of fan:

> On lui mettoit à la main droite (à Henri III) un instrument qui s'estendait et se replioit en y donnant seulement un coup de doigt que nous appelons ici un esvantail: il estoit d'un velin aussi delicatement découpé qu'il estoit possible, avec de la dentelle à l'entour de pareille étoffe. Il estoit grand, car cela devoit servir comme d'un parasol pour se conserver une hâsle, et pour donner quelque refraichissement, à ce teint délicat... Tous ceux que je pus voir aux autres chambres en avaient un aussi de mesme étoffe, ou de taffetas avec de la dentelle d'or et d'argent à l'entour.

There is also a *découpé* paper fan with ivory sticks in the collection of Her Majesty the Queen, which is thought to have belonged to King Charles I.

A taffeta fan of the period, believed to have belonged to Mary, Queen of Scots, survives in the National Museum of Antiquities for Scotland, but it is a cockade fan with sliding tortoiseshell sticks, the leaf of bands of green, yellow and brown silk and silver lace. There is a rectangular hand-screen mount of pierced parchment (Italian, c.1570–80) illustrated in Seligman and Hughes, *Domestic Needlework* (plate 48), from the collection of Albert Figdor.

Right: Fans sold at Christie's South Kensington in February 1988, including: a scene of a banquet with George II and Queen Caroline in Westminster Hall, English, 11ins (28cms), 1727, £3740 ($6171); a fan with ivory sticks, its leaf a trompe l'oeil painted with, amongst other objects, a rolled up print by Francesco Bartolozzi, an invitation to the Ball at the Mansion House, playing cards, a folding rule, a lottery card, and a bank draft for ten pounds, English, 10½ins (27cms), 1775, £5280 ($8712); an articulated fan, the silk leaf painted with a shepherd and shepherdess, the ivory sticks carved, pierced, silvered, and gilt, the guardsticks set with articulated vignettes of two men sawing a plank and a girl picking flowers, French, 10½ins (27cms), c.1770, £1320 ($2178); and a fan with tortoiseshell sticks carved with putti, its leaf depicting a farmyard accident, 11½ins (29cms), c.1880, £495 ($817).

14

By the second half of the seventeenth century most fans were about 12 in. long, with a wide span of almost 180°, painted kid leaves and either ivory or tortoiseshell sticks, more or less touching, simply carved and pierced, and just tapering gently from the tip to the pivot. However, two fine examples of mica fans of this period are known; these have a slightly smaller span. One was sold at Christie's in 1969 and a similar one is in the Messel Collection. It is amazing that an object composed of so many small segments should have survived so long. They were made from many painted panes of mica joined by painted strips of paper to the sticks. This all shows how carefully designed a fan has to be: not only does it have to look decorative, it has to be practical to use, perfectly balanced so that it is comfortable to hold, and to open and close. It is, of course, the pleated paper fan that is the most conventional.

Below: Another arrangement of the fine group of fans sold at Christie's South Kensington in November 1989, including: middle far right, a fan with a chickenskin leaf and ivory sticks, painted, after Canaletto, with the Piazza di San Marco, Venice, c.12ins (30cms), £3250 ($5808); middle near left, shown closed, a Japanese ivory brisé fan decorated by shubayama – work, 10ins (25cms), late 19th century, £2090 ($3448); and top left of the right hand group, a peach-coloured silk-leafed fan with an overlay of Brussels lace with gauze insertions painted with butterflies, its sticks made of ivory, 13ins (33cms), c.1880, £286 ($472).

In 1765, Diderot published *Recueil de Planches sur les Sciences, les Arts Libéraux et les Arts Mécaniques* to illustrate his encyclopaedia. The four plates on fanmaking illustrate clearly the routine in a fan workshop. It is interesting to note that all the workers illustrated are women. A very large proportion of women were employed in the fan trade in the eighteenth century.

The first plate shows a large well-lit workroom with five women preparing the paper for the leaves. The first girl stands at a large L-shaped table with a pile of sheets of paper, still rectangular, a bowl of glue and a sponge. She glues two sheets together which the second girl then stretches on a fan-shaped frame. The third then hangs the frame from the rafters. The fourth takes the frames down when dry and removes the papers from them, piles the frames up and hands the papers to a fifth girl who trims them to a fan shape. The stages are all numbered and explained in a key and below them are scale drawings of the equipment: the frames or *rondes*, and the *sonde* or probe which will be used later, and which resembles a small billiard cue about 14in (35.5cm) long. There is a 'stone' and mallet, used for gilding fan leaves. Mrs. Alexander of the Fan Museum points out that the stone is probably an agate, which is used by gilders to smooth and brighten gold leaf.

The second plate illustrates the fan painter, also female. She sits alone in a grander room at a large bureau with the fan leaf flat in front of her. On an easel in front stands the picture she is copying.

Plates 3 and 4 show the next stage, the mounting of the fans—a very exacting task. Two women sit at large tables with large cupboards behind them to store the finished fans in. The first woman works out where the creases are to go. To do so, she lays the leaf on a large walnut board, etched with radiating grooves. She has more than one of these boards and selects the most suitable for the eventual span of the fan, arranging it carefully so that creases do not cut across the design at the wrong or ugly places. Then with a tool called a 'jeton' – 1in (2.5cm) in diameter, with or without a handle and made of silver or copper—she creases the fan by running the *jeton* along the paper above the grooves. Next the centre semi-circle of the leaf is cut out and the leaf is pleated like a concertina along the lines already marked. Then the second girl with the probe or *sonde* eases apart the two layers of paper between the folds at the bottom edge of the fan to insert the sticks, or rather the plain extensions to the sticks that one never sees on a finished fan. Although it is possible the sticks were made in another part of the same workshop, it is equally possible that they were imported from as far away as China. Before

Above: Early 18th century. Here the leaf is painted with musique champêtre, *the figures dressed in Turkish style, and the borders are decorated with spangles. The ivory sticks are carved with figures dancing and with buildings, and the guardsticks are decorated with* pierrot *and* piqué *in silver. 11ins (28cms), sold at Christie's South Kensington in February 1990 for £308 ($508).*

mounting the leaf on the sticks, the leaf is folded up tightly and trimmed at either end. The sticks are then slid into place and a band of paper is bound round the outer edge of the leaf.

These, therefore, are all the manoeuvres which have to be performed in preparing a folding fan. (Another good illustration of the tools used by the fanmaker is the trade card of the Dutch fanmaker Frans Beenevelt de Jonge.) Not all fans had double leaves: late seventeenth- and early eighteenth-century Italian and Flemish fans seldom do. Fans with single leaves are often said to be 'monté à l'Anglais', probably because, as paper taxes were high, the English had to be even more economical than others.

The fan illustrated by Diderot was just a straightforward simple fan. As we shall see now, they were not always so, and often countless other details had to be added.

By the beginning of the eighteenth century the sticks had begun to be decorated. Both ivory and tortoiseshell sticks were often *piqué* with small studs or strips of silver

Above: German, probably Augsburg, c.1730. This charming scene shows peasants resting after the vintage. The ivory sticks are printed with flowers and the guardsticks clouté *with mother-of-pearl and tortoiseshell. The pivot has a white metal handle. 11½ins (29cms), sold at Christie's South Kensington in January 1980 for £150 ($248).*

or gold. The finest *piqué* is said to have been done in Paris and Naples, but it was also practised in England and Holland. Late in the eighteenth century *posé d'or* was occasionally used, that is, tortoiseshell inlaid with chased gold designs. At this point some sticks begin to be *clouté* with mother-of-pearl, often in the form of Harlequin or other figures from the Commedia dell'Arte, and sometimes with birds. Another combination was found on some Italian mid-eighteenth-century fans: the ivory sticks have mother-of-pearl guards *clouté* with tortoiseshell. Another type of stick to be found in the first half of the eighteenth century is of stained ivory; sometimes only the guardsticks are stained. One Italian fan sold at Christie's had its ivory guardsticks stained red with veins in a jigsaw-seaweed motif to resemble marble.

Most of these sticks from the first half of the eighteenth century are narrower and trimmer than those of the previous century. They now tend to have a slight neck and *gorge*—that is, the guardsticks taper gently from the tip to where the mount ends and then bulge slightly to form curved shoulders and taper again to the pivot. It is normal for the sticks to spread out again at the pivot to form a handle.

By about 1740 the handle is sometimes carved to form a shell or a sheaf of corn or some other decorative object. One Dutch fan formed a bird's head. Then and also later in the century it was quite normal for the guardsticks to be spliced with another material at the handle, such as ivory with tortoiseshell or mother-of-pearl, or wood with bone. This is both strengthening and decorative. Occasionally the guardsticks are made of a different material from the other sticks: for example, tortoiseshell guardsticks and ivory sticks. The normal span of a fan of this period is about 120°, whereas late seventeenth-century fans were almost 180°. Again, the sticks still touch when fully open to form a continuous surface.

Where were all these sticks made? It is still not fully known. Most ivory is thought to have been worked in Paris, then towards the end of the eighteenth century in Dieppe, but since many of the craftsmen were Protestants, some must have fled to London and Holland after the revocation of the Edict of Nantes (see Chapter 4). Certainly many of the members of the Fanmakers Company describe themselves as stickmakers.

Throughout the century many sticks were imported from China. Most of the trade with China was by way of France and England. Holland is not thought to have imported many fans from China—individual sailors may have brought them back, and in 1760 the Dutch East India Company is known to have sent a sample or samples to Canton for copying; this was a commercial failure, but Dr van Eeghen has suggested that the subject of one

sample was Perseus and Andromeda, as a number of Chinese versions of a form of this subject have turned up in Holland. At this stage it is still too early to know, but probably most European sticks were made in France, England and the simpler sticks in Holland, and possibly some of the *piqué* sticks in Italy.

By mid-century the guardsticks are more decorated—a superb damaged fan (C.S.K., 23 October 1979) had sticks decorated with mica, hardstones and chased silver—others are composed of ivory, or ivory and mother-of-pearl, carved and backed with coloured tinsels or decorated with straw-work.

In the middle of the century, the shoulders of the guardsticks are more often squared off, not rounded as before. However, they are now often arranged the opposite way, that is, the upper half of the guardstick is broader than the lower half but it is also squared off. The sticks are now often spaced out when open and most fans of this period (1750–80) have a wide span, almost 180° or even over 180° or *à grand vol*. During this Rococo period sticks cease to be carved all alike and they are carved with great movement and gusto, with figures, flowers and rocaille, the pattern moving across several sticks.

From the end of this period date the *battoire* sticks. These are, again, ornamental sticks, spread out and with large decorative pierced medallions in the centre. The medallions are sometimes in the shape of tennis rackets, or of guitars. *Battoire* sticks are always thought to have been much favoured by the Spanish market; certainly there

Above: Spanish, or French-made for the Spanish market, c.1760. A fine fan, the leaf is painted with a lady drinking chocolate. The ivory Battoire sticks are carved and gilt with the arms of Spain. Provenance the Marchioness of Bristol; the Viscountess Harcourt, HRH the Princess Royal; now in the collection of the Rt Hon the Earl of Harewood.

were a number of examples in the 1920 Madrid Exhibition.

Most sticks are either entirely of ivory or of mother-of-pearl. They are usually elaborately carved and pierced with figures and flowers. On the more important fans the carving normally echoes the theme of the leaf. For special commissions, both the sticks and the leaf would have been specially designed and ordered. This is a useful point, as a great number of fans have been remounted on later sticks, or vice versa, because either the sticks or the mount have been damaged. Many sticks, both mother-of-pearl and ivory, are backed with a thin layer of mother-of-pearl. The guardsticks often have a thin layer sandwiched between their two outer layers; it glistens through the carved pierced sections of the top layer as it does when backing the sticks. Both the ivory and mother-of-pearl sticks are usually silvered and gilt and often painted as well. Sometimes they are even decorated with straw-work which can be stained various colours. The sticks are also *clouté* with mother-of-pearl. Silver and gold *piqué*-work no longer seems to occur at this period. At this period the sticks are also backed with coloured tinsels or coloured metallic paper, mainly strawberry pink, or occasionally

Above: The silk leaf of this fan (which is slightly worn at the folds) is painted with a blue urn containing flowers made of straw-work, with birds fashioned in feather-work perched in flowering trees; all is set within a border of spangles. The sticks are lacquered in red and gold with chinoiserie. *11ins (27cms), c.1770.*

with mica. Sticks of tortoiseshell and of horn also occur occasionally, as do stained wooden sticks, although they are more frequently seen in the last two decades of the century. At this time too, one sometimes finds sticks of alternating materials or contrasting colours—wood and ivory or stained and non-stained ivory. But this happened much more frequently on mid-nineteenth-century Chinese export fans, where one might find three contrasting materials, for example, mother-of-pearl, ivory and cloisonné enamel or silver filigree.

From the 1760s date the sticks called 'Pagoda', by Woolliscroft Rhead. They are of pierced ivory carved to resemble bundles of rods. No. 119 in the 1920 Madrid Exhibition was a fine example. Each stick is divided in two lengthwise, joined only at intervals, and so the closed fan does resemble a bundle of rods, not really very Chinese in style, but more like some chairs of the period with legs called Pagoda legs; the chapter on chinoiserie in *Fans from the East* mentioned similar designs in Charles Manwaring's *Designs for Furniture* (1760). A set of these chairs was sold at Christie's in 1969.

By about 1780, sticks became simpler again. Most fans span about 120°. Certainly many of the printed fans have straight narrow plain wooden sticks with a rectangular projection at the tip where they join the mount; they are normally 11in (28cm) long. A number of European fans are mounted on Chinese sticks, sometimes ivory, sometimes wood, and decorated with fretwork. Indeed, the

main decoration on late eighteenth-century sticks is this pierced work, executed both in Dieppe and in China, and perhaps in England too. Some of the fans have plain sticks with just pierced work on the upper, broader part of the guardsticks. On some fans the sticks are now touching again to form a solid surface, possibly because it was at this stage that brisé fans returned to fashion. Some guardsticks were decorated with cut-steel beads.

A superb fan illustrated in *Fans from the East* (plate 16) has guards set with Jasperware plaques. A brisé fan painted with vignettes, possibly by Angelica Kauffmann, had guardsticks set with enamelled portrait miniatures. The sticks, as in some of the more important fans, are not straight but slightly tapered and slightly shaped. These elaborately decorated guardsticks are an advance guard of some of the very elaborate trimmings to be found on early nineteenth-century French printed fans.

Between about 1815 and 1825, French printed fans were normally 8½in (21.5cm) long with a span of about 120°. Their sticks are touching, usually of ivory or bone, shaped in a rather angular way, pierced and *clouté* with steel, sequins or mother-of-pearl. The guardsticks are often very elaborate and can be set with mother-of-pearl plaques, coral, opals and gilt metal plaques. An example sold at C.S.K. was set with enamelled clockfaces.

Above right: A silk-leafed fan painted with figures, its reserves embroidered with gold thread and spangles. The mother-of-pearl sticks are carved, gilt, and silvered. 11ins (28cms), sold at Christie's South Kensington in January 1980 for £340 ($561).

Right: French, c.1760. A farmyard scene is delicately depicted on the leaf, and the ivory sticks are carved and painted. The fan rests in a 20th-century glazed case.

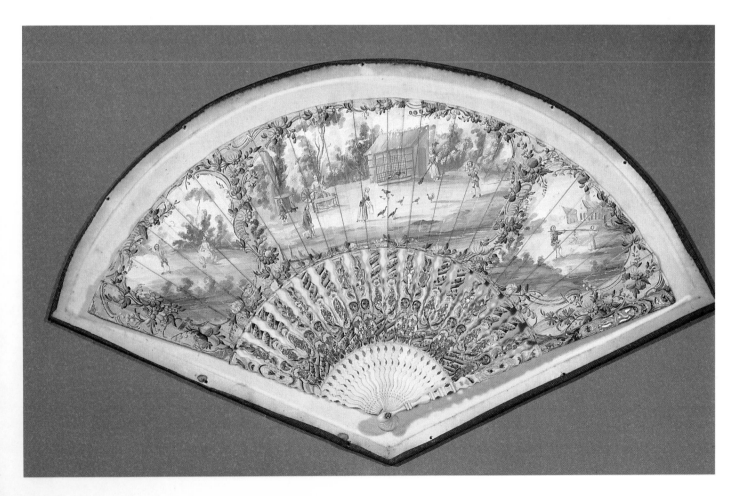

Right: c.1860. An elaborate handle
hangs below guardsticks set with
porcelain plaques, with putti within
enamelled gilt metal mounts also set
with two mirrors. The mother-of-pearl
sticks are carved, pierced, and set with
a central vignette of a hand-coloured
lithograph of figures in 18th-century
dress in a park: the narrow outer leaf
is another hand-coloured lithograph,
depicting a 17th-century court scene.

Fans remain small whilst dresses remain thin and hoop-less, getting bigger in the 1830s and 1850s when skirts grow wider and pockets are again larger.

By the mid-nineteenth century the finest fans usually have mother-of-pearl sticks. Again, they are touching but the fans tend to measure 10in (25cm), and have a span of nearly 180°. The sticks are often very wide, with rounded shoulders tapering to the pivot. The tip of the guardsticks is normally fairly straight and the same length as the mount, which is often quite short, sometimes only about a third of the total length of the fan. They are elaborately carved, pierced, gilt and backed with mother-of-pearl as in the eighteenth century.

The less expensive sticks are of ivory, bone or lacquered wood. Some sticks are even decorated with lithographic scenes. At this period one begins to find gilt metal handles attached to the pivot, often with elaborate silk tassels suspended from them. One does come across eighteenth-century fans with handles, but almost all have been added in the nineteenth century by M. Duvelleroy or some other *éventailliste*. Dr. van Eeghen tells me that her father owned a design for an eighteenth-century fan handle but that she had never come across an example. In fact, one was sold at C.S.K. on 31 January 1980, and a handful have been seen since.

Between 1860 and 1870 sticks grow a little longer, a little narrower and with slightly more angular shoulders. They are also plainer, normally bone with a slight pierced pattern. The expensive lace fans tend to have mother-of-pearl or tortoiseshell sticks. By the 1880s and 1890s sticks

Right: c.1890. A lemon-yellow
gauze-leafed fan with a light spring
ambience. It is trimmed with ribbons
which are not continuously sewn, but
loosely tacked so that they form rosettes
when the fan is closed. On the gauze
itself are delicately painted white and
blue daisies. The sticks are wooden,
carved, and painted. 14ins (36cms).

Above: c.1880. The rich amber-coloured tortoiseshell sticks contrast magnificently with the leaf of gossamer-thin decorated black lace. On the sticks are carved the initials UB with, intriguingly, a coronet above. 13ins (33cms).

are long, about 12in (30.5cm). They tend to be narrower, spaced out and tapering from the tips to the pivot. The ends of the guards often have a little carved floral decoration echoed round the tips of the sticks. They no longer have shoulders. Many are of wood, lacquered white, grey or black.

There are a number of fans of that period which have rather light, shaped and curved sticks. In the twentieth century, the printed fans normally have plain wooden sticks. The feather fans usually have mother-of-pearl sticks if white or coloured, and tortoiseshell if black. The handles of the ostrich plumes popular in the 1920s were normally of stained ivory. In September 1922 the magazine *Art, Goût, Beauté* mentions long heavy tortoiseshell brisé fans with wide sticks as being fashionable.

Now to revert to the mounts. Until about 1780 they were normally of vellum or paper. The more valuable were sometimes embellished with many of the same materials as the sticks, namely mica in the late seventeenth century and the mid-eighteenth century, and straw-work and mother-of-pearl in the mid-eighteenth century. To this one can add spangles in the early eighteenth century, feathers, occasionally butterfly wings, and applied silks in the mid-eighteenth century. A little later, when many of the leaves were of silk, they were trimmed with gold braid, sequins, and spangles again, and even occasionally had painted ivory faces, as did Cantonese fans of the nineteenth century.

Some fans were of silvered paper, particularly from about the mid-eighteenth century to the 1780s. This was often decorated with cut-paper or *découpé* work. Others were decorated with bands of pressed silver paper. The fans with spy holes had insertions of gut.

Occasionally one finds eighteenth-century lace fans (there is a fine example at Waddesdon). Sometimes they are set with painted vignettes; some French fans of the 1760s were of darned muslin. But it was not until the 1860s that lace was used in quantity on fans, and some very fine examples survive. White lace was mounted on mother-of-pearl sticks, black lace on tortoiseshell sticks.

The mounts were made 'à disposition', to the correct shape for the fan. Obviously examples of varying quality survive. The lace used is mainly Brussels, Honiton and Chantilly. Some of the finest examples are worked with scenes. A fine fan sold at Sotheby's Belgravia in May 1979 was worked with scenes from *Don Quixote*.

There are also some rare examples of seventeenth- and eighteenth-century embroidered fans, and there is a mid-eighteenth-century Dresden-work fan leaf in the Cooper-Hewitt Museum, New York.

In G.S. Seligman and T. Hughes, *Domestic Needlework*, several examples of embroidered fans are illustrated: colour plate XVIIB: a fan screen embroidered with a vignette of a pelican in its piety with a silver border, French, late seventeenth-century; plates 47A, B and C: a pair of

23

Top: English, c.1760. A Court fan, the paper leaf painted with a queen and a hero, the reverse with a lady and Cupid; the pierced ivory sticks set with rose diamonds, sapphires, rubies and emeralds. 12ins (30cms); in original sharkskin fitted box.

Above: English, c.1720. An ornate border of spangles, with eagles, flowers, and putti the devices, surrounds an otherwise quite plain fan, consisting of a leaf painted with a simple rustic scene of a shepherd and shepherdess; undecorated ivory sticks; and tortoiseshell guardsticks. 10ins (25cms).

needlework handscreens of upturned pear-shape worked in coloured wools with chinoiserie figures like chair covers of the period, English, c.1720, from the Percival Griffiths Collection, and a single hand-screen, similar; plate 48c: another hand-screen also worked in petit-point, with a shepherdess, English, c.1740. There are also two interesting examples from later in the century. Colour plate XVIIA: an unmounted fan leaf of gauze embroidered with David and the head of Goliath and vignettes of fountains, against an unworked ground, probably Italian, c.1775–1800; plate 46A: a linen leaf embroidered in silver and coloured silks with three oval landscapes and delicate floral reserves, Italian, c.1780–1800.

In the 1760s sequins and braid were often embroidered round painted vignettes on silk fan leaves. Early in the nineteenth century gold, lace and sequins became more and more prominent on French fans, including those for the Spanish market. These fans, profusely embroidered with sequins, were revived in a similar style in the early twentieth century. This time, however, the ground was usually completely covered with sequins, no longer normally gold but of various bright colours, sometimes with mother-of-pearl sticks stained to match the sequins.

From the 1840s one finds pairs of hand-screens embroidered in beadwork like the footstools of the period. They are often circular, trimmed with fringes and with turned wooden handles.

FAN BOXES

One of the reasons many fans have survived in such fine condition is that they have sometimes been stored in boxes, occasionally their original boxes. Many of the English eighteenth-century and French early nineteenth-century

fan boxes retain the fanmaker's label. It is, therefore, beginning to be possible to recognize some fanmakers' boxes. Clarke favoured boxes of hexagonal section covered in green paper and overpainted. Most eighteenth-century boxes to survive are of papier mâché lined with printers' waste. The English boxes are all long canisters with a removable cap either hexagonal or elliptical. They are covered in pink, white, blue, red and orange figured paper; some English boxes are also covered in French paper. Dutch boxes are long narrow rectangular boxes with lids covered in paper of various colours figured with stencilled motifs, often in a simple cross pattern. Boxes without labels may indicate that the fans were sold in other shops. In 1988 Sotheby's Sussex sold a fan box with the label of Bath Milliners.

Tapering rectangular boxes with hooks and hinges covered in shagreen have also survived, and occasionally French or English late eighteenth-century oval tubular boxes of red morocco. The collector must beware of oval paper tubular boxes covered with imitation scarlet leather; these date from the 1840s, as can be seen from the typography of the paper used in the papier mâché. A box, formerly in the collection of Lady Charlotte Schreiber and sold at Sotheby's Belgravia in May 1979, contained an Act of Parliament of 8 Victoria (i.e. 1845).

In the nineteenth century most of the boxes that survive are of the same type as Dutch eighteenth-century ones, but they are often hinged. Regency boxes, however, are

Above: A collection of 18th-century and early 19th-century fan boxes, including one mid 19th-century example (centre).

Above left: A group from a sale at Christie's South Kensington in June 1987, top to bottom: French, c.1770. The leaf is painted with a court scene, possibly at Naples, and the ivory sticks are pierced and silvered, 11ins (28cms), £605 ($998). English, mid-19th century. An unmounted fan leaf painted with allegories of painting, music, sculpture, and learning with Jupiter in the Campagna, 19ins (48cms). English, c.1760. The silk leaf is painted with three chinoiserie figures in a garden, their clothes of applied silk. The window of a kiosk and a handscreen of mica form spyholes, the sticks are of ivory, 10¼ins (26cms), £715 ($1180). French, mid-18th century. A telescopic fan painted with sprigs of flowers and applied with gold spangles of leaves and flowers, printed Clarke and Co., No. 26 Strand, Invent., 9ins (23cms) £110 ($181).

often rectangular versions of English eighteenth-century boxes. There are also some fine morocco coffin-shaped French fan boxes of this type from about 1810. From the 1840s boxes are mainly covered in printed paper, but sometimes wooden and lined with quilted satin. By the 1880s most of the boxes are now domed and are covered in white, black or pink satin, lined with matching paper, with the fanmaker's label stamped in gold inside the lid. Canton fans have the grander and lacquered boxes, but many of the finest ostrich feather fans by Duvelleroy come in enormous rectangular white satin or white paper fitted boxes.

A number of late nineteenth-century tooled leather boxes also survive. There are also a number of boxes, probably early twentieth century, covered in eighteenth-century and later silks; they often bear Spanish labels.

SEVENTEENTH- AND EIGHTEENTH-CENTURY PAINTED FANS

A painted fan is a memento: it carries a

memory, a story, a plea or a message. As the

eighteenth century progressed, the popularity of

fans continued to increase.

It is unlikely that the collector today will come across fans earlier than the second half of the seventeenth century. However, in the 23 years of specialized fan sales at Christie's, about three dozen fans dating from the last few decades of the seventeenth century have come on the market and so it should be possible for the serious collector to be able to include at least one in a collection.

BAROQUE

So far three types have emerged. The first is a very long fan, about 12in (30.5cm), painted on a coarse kid leaf with a wide span of nearly 180°, and painted all over with a rather loose but well-balanced composition normally incorporating fat putti, in bright colours often enhanced with gold. These fans have rather heavy tortoiseshell sticks with sparse decoration. In fact, the sticks are almost always identical in shape although sometimes of ivory rather than tortoiseshell. They probably come from the same source. There are fan leaves of this type in the Schreiber Collection (in fact, they hover between this and the next type in style—the subjects are similar to the first but they are less elaborate than the second), such as a French fan leaf painted with an allegory of the marriage of Louis XIV. The scene is painted as if planned as a rectangle and cut down to a fan shape, sometimes cutting across arches, which gives the fan a very theatrical air. The ladies-in-waiting are holding fans and cupids are making the bed.

Other examples in the Schreiber Collection include 'The Lovers' Agency'—in a classical building on an island

lovers approach tables covered in green cloth and are presented with cards inscribed 'Congé pour un Amant Constant, etc.'; 'The Toilet', painted with a lady at her toilet in a room steeped and scattered with flowers as her lover enters, and 'The Bride'—again, a very open stage-set-like arrangement containing a large four-poster bed. The young bride is embraced by the groom as attendants undress her. By the door friends receive the priest.

In the Walker Collection there were two further fans of this type but with classical scenes of Diana and Endymion and the chariot of Phoebus. Here the bottom edge is painted with flowers.

In this group of fans there are a number which are much simpler, depicting perhaps just pots of orange trees surmounted by putti and drapes. Some of these have been catalogued simply as North European, others, the more elaborate ones, as French.

The second type of Baroque fan is far superior in quality. It is almost certainly French and depicts very elaborate

Right: A highly attractive group of fans from a sale at Christie's South Kensington in December 1988, with several 17th and 18th-century examples, including: middle left, a fan, the leaf painted with a family in a park, the verso with a lady, the ivory sticks carved and pierced, and painted with chinoiserie scenes, the guardsticks carved with huntsmen carrying shotguns, English, 11ins (30cms), c.1750, £495 ($817); centre, a fan, the leaf painted in the manner of Moucheron with a theatrical scene, probably Mozart's 'Entfuhrung aus den Serail', the ivory sticks pierced and silvered, Dutch, 10ins (25cm), c.1770, £605 ($998); and, top left, a fan, the leaf painted with an elegant shepherd and shepherdess, the verso painted with sprigs of brightly coloured roses, carnations and tulips, the ivory sticks piqué in silver, English, 11ins (30cms), c.1690, £2000 ($3300).

scenes filled with figures—all well painted. Their costumes in particular are superbly handled. As the previous type, they are painted in bodycolour, usually on a dark ground, but on a thinner kid leaf of the type called chickenskin. A fan leaf of this type was sold at Christie's and is now in the Victoria and Albert Museum; there is also an enchanting series in the Musée des Arts Décoratifs and the Musée Carnavalet of market scenes and other everyday Paris views. These all tend to be longer than the fans of the next few decades.

The third type is of mica and much smaller in size than the two previous categories. Several examples of this type are known. One, which was in the Brompas Collection at Christie's in 1969, was painted with peacocks, busts and flowers, and has tortoiseshell sticks. Another is in the Messel Collection. The fans echo the construction of the late sixteenth-century fan described by Bapst. The painted panes of mica are held in place by paper frames. Each slice of mica is painted with a brightly coloured jewel-like miniature, a figure, a bust with elaborate hairstyle, an animal or flowers. Those, too, were probably French.

Since figures in contemporary dress appear in most of the scenes, as usual the best way of dating them is by the fashion of the costumes or hairstyles.

Many more fans survive from the early years of the eighteenth century, and hence there are more to compare and assess. It is possible to sort them into groups but it is still not always, or even often, possible to be certain of their country of origin. The chief way of dating and attributing all painted fans of the seventeenth, eighteenth and nineteenth centuries is by the style of painting, that is, by trying to compare them with watercolours, portrait miniatures and oil paintings that have already been ascribed a country of origin and a date.

There are many things that make this a difficult task. The manufacture of fans was fairly widespread and the trade even more so. There was much importing and exporting of both whole fans and fan parts. Also, many fans were not well painted, but executed by journeymen and hack-painters, and very little about them has been discovered, although gradually gleanings from contemporary local newspapers, registrations at town halls, insurance policies and wills are beginning to unravel a pattern. We have seen that sticks were often made elsewhere from the mount. But there is another equally confusing point. As in oil painting, a fan leaf was not necessarily painted by one person alone. There are several examples where the borders appear to be Chinese and the central design European. Also, in portrait painting, some artists are known to have been specialists and to have painted only the heads whilst one assistant painted the hands and another the drapes or background (for example, Sir Godfrey Kneller always called himself 'face-painter' and for most of his practice employed Capt. Byng to provide compositions and stances, Alexander van Haecken to paint drapes, and sundry dog and cat painters

Below: English, c.1740. A fan with a leaf painted with three classical scenes, the ivory sticks carved, pierced, silvered and gilt. 10ins (25cms).

Above: English, c.1690 (leaf slightly rubbed). A fan with ivory sticks piqué in silver, the leaf painted with an elegant shepherd and shepherdess, the verso with sprigs of brightly coloured roses, tulips, and carnations. 11 ins (28cms), sold at Christie's South Kensington in 1989 for £2200 ($3630).

to paint other relevant parts); this certainly happened with fans. Dr. van Eeghen has found instances where Dutch fan workshops employed artists who specialized in flowers and in faces and were employed just to finish off these parts of fans. The reverse of the fan is almost always painted in a weaker style by a lesser artist.

By the third quarter of the seventeenth century Paris and Versailles were becoming the centre for fashionable and luxury taste, a position Paris has held on and off ever since. In 1673, an Association des Eventaillistes was formed with the King as patron. To qualify as a member, the fanmaker was required to serve four years' apprenticeship, and sometimes to produce a *chef d'œuvre*. The entrance fee was 400 livres. Widows could inherit their husband's membership if they did not remarry. There were 60 founder members. When the Edict of Nantes was revoked in 1685, the Protestant members fled to London and Holland, but by 1753 the Association des Eventaillistes had increased again to 150 members.

In England, too, the trade became organized when in 1709 the Worshipful Company of Fanmakers was granted its Charter by Queen Anne, the last purely trade City company to be formed. There were apparently 200 to 500 people sufficiently involved in the business living in London or Westminster or within a twenty-mile radius at that time; probably some were French refugees. Sadly, the minutes only survive from 1747 with the enrolment of the 39th member. On 1 July 1751, Thomas Coe, stickmaker, Bethnal Green, was admitted as a member. No. 882,

Francis Chassereau, Jr., was admitted as fanmaker 3 November 1755; No. 883, Robert Clarke, 12 September 1755, at Mr. Clarke's in Bell Sauvage Yard, Ludgate Hill; No. 936, Sarah Ashton, 1 February 1770. The Company's motto is 'Arts and Trade Unite'. The Company today is better known for the products it manufactures for the engineering industry.

So, we have on one side the French protecting their trade with the Association des Eventaillistes and, on the other side of the Channel, the Company of Fanmakers.

Whilst England and France were exporting many fans, they were also importing a large number, particularly from the East at the turn of the seventeenth and eighteenth centuries. The Chinese taste was the vogue. European fans from the beginning of the eighteenth century often show influences of the Chinese taste, particularly on brisé fans, which make their appearance in Europe at this time. In Savary de Brulou's *Dictionnaire Universelle de Commerce* (1723) under 'Eventail': 'Les Eventails de la Chine et ceux de l'Angleterre qui les imitent si parfaitement, sont les plus en vogue; Et il faut avouer que les uns ont un si beau lacque, et que les autres cédent aux beaux Eventails de France, ils leur sont au moins préférables par ces deux qualités.' Unfortunately, we do not know to which type

Right: , *c.1720. A fan leaf painted with the Triumph of Venus, dismounted and framed in an English frame, c.1740.*

of English fan he refers, as very few fans surviving from the early eighteenth century have been identified as English. This may, of course, apply to Chinese fan leaves mounted in England, as he admits he cannot tell the difference between Chinese and English fans, or rather that English fans imitate Chinese fans so well.

To return to types that have an identified country of origin: France continued to produce superbly painted fans,

Below: Italian, c.1700. A dismounted fan leaf extended to form a rectangle and set in a velvet frame, painted with a lady at her toilette in a garden by a follower of Pietro da Cortona. Previously from the collection of the late RE Summerfield, 10 × 16ins (25 × 41cms), auctioned at Christie's South Kensington in February 1990 for £770 ($1270).

on paper leaves and on chickenskin, which until about 1735 were still very much High Baroque in style, on dark grounds with mythological or theatrical subjects and pretty flowers on the reverse. The sticks became neater and were often *clouté* with mother-of-pearl plaques. The fans also became a little shorter.

In Italy, the main subjects were classical, and also often taken from famous Roman paintings such as Guido Reni's *Aurora*. Thus they were very suitable gifts to bring back from a Grand Tour. Early in the century they were also Biblical.

These Italian fans are often on very fine chickenskin. A number have been signed and inscribed, which is a great

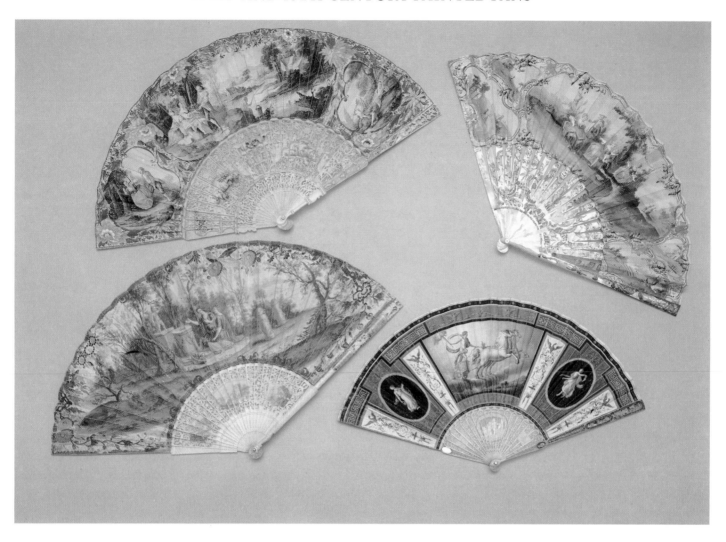

Above: A splendid group of fans from a sale at Christie's South Kensington in February 1988. Top left, the leaf is painted with an angling scene, the sticks are ivory, probably English-made for the Spanish market, 11ins (28cms), c.1740, £990 ($1633); top right, elegant figures in a park, French, 10ins (25cms), c.1760; bottom left, two figures at the Altar of Love, English, c.1740 (later sold in May 1988 for £495/$817); and a fan with a chickenskin leaf painted with Apollo, Italian, 9ins (23cms), c.1790, £715 ($1180).

help in identifying them although it would be of even more use had these painters been recorded elsewhere. For instance, one fan, depicting Diana and attendants resting, is by Leonardo Germo of Rome, 10¾in (27.5cm) long. Fans were not only sold retail to tourists and to the rich Italian nobility, but also exported in quantity, particularly to England. Another fan by Germo, the 'Triumph of Mordecai', was exhibited in the 1870 South Kensington Museum Exhibition and is now in the Victoria and Albert Museum.

An interesting group of three fan leaves was sold at Christie's in their original fanshaped box for export via Leghorn, and two are now in the Victoria and Albert Museum, Department of Prints and Drawings. One is a drawing in black ink; other examples exist in red ink. It is possible that many of these fan leaves were mounted in England on arrival. One was signed Dominico Spinetti, Napolitano, Roma F. The wooden box was inscribed A. Sigri Baracchi and Mucotti Livorno.

Another type that survives occasionally is the early eighteenth-century *découpé* fan of vellum. There is a fine example in the Messel Collection (see Peter Thornton in *Antiques International*, ill. 7). This has a central painted classical cartouche and several small vignettes; the remaining ground is of kid *découpé* to resemble lace. The sticks are more elaborate than those on their French counterparts

at that time; they are pierced and carved and are no longer more or less straight, but have a bulbous shape at their necks. There is a superb German *découpé* fan in the Bayerischen National Museum, Munich, with a vignette signed by Ignaz Preissler, Breslau.

Also at the turn of the seventeenth and eighteenth centuries, brisé fans appear in Europe. As far as has been established, they appear to have been produced only in France and in Holland. The French versions are the type that collectors of fans describe as Vernis Martin—a term which owes its name to the shiny lacquer-like varnish applied to the painting. They vary in quality but are often superb. They are smaller than other fans of the period and are painted with both classical and theatrical scenes, using the sticks as if there were a separate mount. At the *gorge* there is sometimes a narrow border, usually painted with small chinoiserie scenes. The other type, probably Dutch, is more lightly painted on a central medallion and the reserves are pierced to produce Baroque scrollwork.

Right: English, early 18th century. A fine fan painted with Blind Man's Buff, after Francis Hayman. The ivory sticks are inlaid with mother-of-pearl. This was sold at Christie's South Kensington in 1980 for £1000 ($1650).

Above: Early 18th century. A beautiful ivory brisé fan painted and lacquered, showing the puppet show episode from Don Quixote in great detail. The reserve contains a painting of ladies in a garden, seated in front of a fountain. Sold for £1000 ($1650 in 1980 at Christie's South Kensington.

Above: French, c.1760. The leaf is painted with a shaped vignette of a classical scene, and two smaller vignettes of a hunting scene and ferry boats and shipping in a port. The reserves are painted with delicate flowers, birds, fruit, and butterflies, and the verso with figures watching shipping from a cliff top. The mother-of-pearl sticks are carved, pierced, and painted with red and green berries, the verso painted with pink and blue rocaille. *11ins (28cms), sold at Christie's South Kensington in February 1987 for £858 ($1415).*

The first type, in particular, was very popular in the second half of the nineteenth century, and the collector may be confused by reproductions or heavy restoration, especially as a handful of undecorated fans still survive.

ROCOCO

By 1735, there is a preference for lighter, more Rococo styles, and painted and printed leaves are preferred to the brisé. In France subjects are now often pastoral and romantic. The leaves now span out to almost a semicircle, having been about 120° at the beginning of the century. They are now always painted on light grounds and the sticks have become a much more important entity. These are predominantly of ivory or mother-of-pearl and elaborately carved.

Although designs on leaves cover the whole mount, it is at this stage that the designs begin to be broken up into vignettes. There is a series of fans, probably Flemish, where the leaf is divided asymmetrically, often by a border of rocaille decorated with glitter, into two compartments differing greatly in size.

In England the topographical fan appeared about 1740. Some of these have been identified by John Harris as being by Thomas Robins, including one sold at Christie's in 1972 of elegant company visiting Ralph Allen's stone quarry at Prior Park, Bath. On this fan the reserves are decorated with a geometrical pattern. Other examples, possibly by Robins, include a view of Belvoir Castle (de Vere Green Collection, colour plate 26). Another example of a topographical fan, a little later, is the bright and colourful view of Covent Garden Market from the Baldwin Collection; it is rather crudely drawn, but this is typical of many English fans. Another interesting example is the view of Cliveden, possibly by Goupy; and at Wilton House near Salisbury there is a fan painted with a copy of a view of the house by Richard Wilson.

Above: A lacquered ivory brisé *fan, painted, after Teniers, with a Dutch scene, late 19th century in 18th-century style, featuring men playing bowls, watched by another figure smoking a pipe. The reserve is ornately patterned, containing a vignette, and the guardstick carries panels with paintings of landscape scenes, an urn with flowers, and a figure holding a wand.*

In Holland, at the time, leaves were painted with pastoral scenes, often very finely drawn and predominantly in tones of pale greens and blue. These fans were fairly small but later, about 1760, they span wider. In Venice, fans remained large, *à grand vol.*

During the 1760s the shape and design of fans change. Far more fans survive from this period and they are now

Left: English, c.1775. A trompe l'oeil fan, with ivory sticks. The left hand portrait is of Elizabeth Gunning, successively Duchess of Hamilton and Argyle; the style is after Francis Cotes R.A. 11ins (28cms), sold at Christie's South Kensington in January 1980 for £700 ($1155).

Right: English or Dutch, c.1740. The leaf is painted with a scene of figures in a landscape; the ivory sticks are stained green and lacquered in gold with chinoiserie.

Left: English, c.1752 (slightly damaged). A topographical fan, its leaf painted with a view of Ranelagh Gardens, Chelsea, with the Chinese pavilion in the foreground, the verso painted with chinoiserie figures: it is after the lost painting by Canaletto. The ivory sticks are carved and pierced with vases of flowers, and gilt, and painted with blue flowers and ribbons. 10ins (25cms).

normally arranged with one large and two smaller vignettes. The finest French fans are exquisitely drawn and tend to be more sombre in colour than those of other European countries. The cheaper French fans are usually painted with dumpy figures with round pink faces.

Typical French fan subjects of the time were pretty groups of lovers, sacrifices to Hymen, families and shepherdesses. A fine example commemorating the marriage of the Dauphin and Marie Antoinette in 1770 was sold at c.s.k. in 1978. This fan can be traced back as previously sold at the Hôtel Drouot, Paris, 13 April 1897 (lot 1, as the property of Madame x).

In Holland a similar arrangement of three vignettes is deployed but leaving much more blank background. One finds a series of fans commemorating Zoutman, the Dutch commander at the Dogger Bank in 1781. Most Dutch fans, however, have Biblical and pastoral subjects. Fans were used in church, where respectable women prayed behind their fans, whilst men used their hats. The Biblical fans were mainly of Old Testament subjects, those of the New Testament being considered idolatrous by Calvinists. By the early nineteenth century the custom of taking fans to church began to die out. Dr. van Eeghen, in her article in *The Bridge* (no. 8, 1966), says that young women took any fan to church, although it is not known if they went as far as Englishwomen of an earlier generation, who took to church fans painted with 'naked Cupids and women almost so', to the horror of *Lady's Magazine* in March 1776.

Above: English, c.1760. A rare fan, the leaf painted with an extraordinarily exaggerated hair style, including military fortifications: further military scenes are depicted within vignettes either side, and the whole is set off by a reserve painted with butterflies, and carved and pierced ivory sticks.

Above: The leaf is Italian, the sticks English, c.1740. The leaf is painted with King David and a procession bearing the Ark of the Covenant. The sticks are carved and pierced ivory, decorated with chinoiserie. 11½ins (29cms).

A fashion for *trompe l'œil* fans emerges in the 1760s. These are painted with a medley of drawings, pictures, playing cards, jewels and lace, all lying on a marble table or a silk table-cover, and are often dated and signed, sometimes by Italians and Germans; the signatures are normally disguised by being on one of the paintings such as one sold

Above: Italian, 1760; the sticks are possibly French. A trompe l'oeil fan of prints lying on materials with mother-of-pearl sticks silvered and gilded. The engraving is signed Angelar Albonesi Romanus. *Sold at Christie's South Kensington in March 1985.*

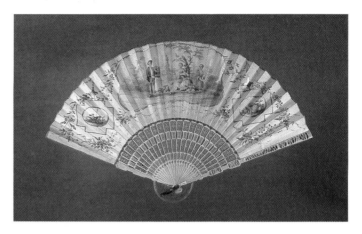

Above: Dutch, c.1760. The leaf is decorated with blue, yellow, and pink flower devices. The central vignette shows two ladies and a gentleman in a pastoral setting, the man holding a bird cage. The two smaller vignettes contain still lifes of fruit and flowers. The Chinese sticks were probably imported from England.

at c.s.k., which was signed 'Ioh Georg Hertel, excud. Aug. V', from Augsburg in about 1760. There is also an example in the Castle Museum, York; there was a German example in the Baldwin Collection; and c.s.k. sold a Dutch and a Flemish example in 1979.

A number of pretty examples with much lace in them are to be found in the Madrid Exhibition of 1920; they were probably French or Italian, made for export to Spain. Late examples of the genre are Assignat fans, with leaves painted or printed with an assortment of bonds used in Revolutionary France. Some *trompe l'œil* fans were also certainly English. Perhaps it is true to say that *trompe l'œil* style is universal. There was a charming earlier Italian example in a Christie's sale in 1969, with the *trompe l'œil* used merely as reserves.

NEO-CLASSICAL

In Italy, from the 1770s until the turn of the century, most fans were painted with views of classical ruins, with reserves with Pompeian decoration. There were also dramatic views of Vesuvius erupting and of the bay of Naples. They are often inscribed, sometimes signed and dated, often with the city of the artist added, such as Camillo Buti, Rome.

At this period, there were still a number of classical scenes, in Pompeian reserves. As Mrs Alexander put it so charmingly in a lecture, 'The figures are now draped and much closer to the antique'—this was a result of all the excavations and finds of those Englishmen, the Dilettanti, and others on their Grand Tours studying the antique.

In France, from the last quarter of the eighteenth

Above: c.1700. The leaf of this fan is painted with a classical scene of a hero leading a Queen onto a ship; the verso is painted with chinoiserie with three vignettes mainly in blue and white. The ivory sticks are painted and gilt with chinoiserie. One of the guardsticks is missing, the other loose. 9ins (23cms).

Left: French, c.1770. The silk leaf is painted with three vignettes of an elegant lady buying shuttlecocks for her child from a pedlar; a wheelbarrow containing baskets of pink-flowered plants; and a farmyard scene with chickens walking among sheaves of corn. The sticks are ivory, carved, pierced, and gilt, and the reserves are embroidered with spangles. 10½ins (27cms), sold at Christie's South Kensington in May 1989 for £308 ($508).

century until the Revolution, there are a number of very pretty fans, often painted on silk. Some are painted with figures in a park or a garden, dressed at the height of fashion, sometimes playing at shepherdesses as Marie Antoinette loved to do at Le Petit Trianon. A number of fans from this period commemorate balloon ascents; they are very much sought after. But from this moment, until the Restoration, France did not produce many more painted fans. Cheap printed fans with political slogans were popular during revolutionary times. France then, soon after the Revolution, led the field of the finer printed fans at the turn of the century. And under Napoleon small silk fans, often decorated with spangles, were popular.

In England, the brisé fan returned to favour, longer than those of the early eighteenth century and still as narrow, often embellished with printed decorations but sometimes with very well-painted vignettes. It is probable that some of these ivory brisé fans were carved in Dieppe: excellent brisé fans were being produced by the renewed Dieppe industry at that time and an example was presented to Marie Antoinette.

Having broadly covered France, England, Italy and Holland from whence most European fans emanated at the time, it now remains for us to try and sort out Spain, Portugal, Germany, Switzerland and North European and Scandinavian countries such as Sweden.

Fan collectors and enthusiasts are still trying to group fans into countries of origin. It is very hard to do so since, as we have already seen, sticks and leaves were often

Left: Mid-18th century. The leaf is painted with three vignettes, an elegant couple in riding dress with a hound; a fisherman by a lock gate; and quayside with a church and a fort in the distance. One fishing boat flies the Dutch flag. The reserves are painted with birds and blossom in grey against a brown ground, the verso with blossom, the ivory sticks carved, pierced, and painted with flowers. 10ins (25cms), sold at Christie's South Kensington in February 1990 for £462 ($762).

imported from different countries. What one can do, however, is to pore over old catalogues of exhibitions in the nineteenth century and early in this century, and catalogues of collections, and work out where most of a certain type were found; it is then probable that they originated there or were commissioned for that country. There was a magnificent exhibition in Madrid in 1920 with 491 exhibits, from which one can judge which types of fans the Spanish favoured.

No. 133, for instance, is a *trompe l'œil* fan with two oval portraits and other pictures strewn amongst playing cards; the Ace is inscribed 'Real Fabris Madrid 1757', but the fan is painted in the French manner. It is therefore probably French for the Spanish market. No. 146 is a printed calendar fan, in Spanish, probably English for the Spanish market.

One has to remember the strong political ties Spain had with both Naples and the Vatican at that time. Exhibits Nos 109 and 110 were two fan leaves lent by the Spanish National Library painted with views of the triumphal square in Naples for the return of Don Carlos Borbón, Infante de España, from Sicily. They were designed by Francisco de la Vega (a Spaniard), and executed by Cayetana Sardi, Rome. There are two other fans on the same subject after de la Vega in the Schreiber Collection: the entry of Charles, King of the Two Sicilies (Don Carlos Borbón to the Spanish), into Naples (1734), painted by Gaetano Pichini, Rome, and the review at Gaeta, painted by Leonardo Egiarman, Flamenco ('the Fleming'). They are drawn with a pen in bistre and washed with Indian ink; ornaments are composed of fleur-de-lys similarly drawn on a ground slightly coloured red.

The Spaniards seem to have favoured plenty of portraits and elaborate sticks. In No. 88 in the Madrid Exhibition, the leaf is printed with lovers and two miniatures of a couple and the tops of the sticks are carved with beautiful portraits, probably five versions of the same couple (*c.*1760; from the collection of Queen Victoria Eugenia of Spain). The sticks of No. 159 are set with eighteen portrait miniatures.

No. 67 in the exhibition is an unusual fan with carved and pierced serpentine sticks; the painted leaf, depicting Alexander the Great, has scalloped edges rather like webbed feet. It dates from about 1770, and is similar to a German one sold at C.S.K.

Right: A very handsome group of fans from The Shirer Collection, at Christie's South Kensington in December 1988, including: a depiction of The Rape of Europa, the verso with sprays and sprigs of flowers, with ivory sticks carved and pierced with chinoiserie. 11¼ins (29cms), c.1760, £550 ($907); a fan with ivory sticks, carved and pierced with putti and musical instruments, its reserves painted with chinoiserie figures (their faces of ivory, their clothes of silk, and decorated with mother-of-pearl), and its leaf decorated with a hand-coloured oval stipple engraving of Apollo resting his horses, French, 11ins (28cms), c.1780, £1650 ($2722); and an Anglo-Flemish fan, the leaf depicting scenes of courtship with amoretti, the verso with a lady and putto within a garland of flowers, the ivory sticks carved with dolphins and painted with vignettes of buildings in red, blue, and purple monochrome, and with birds and fruit. 11ins (28cms), c.1750.

No. 75 has sticks with two Chinese figures in the handle; the sides of the sticks are elaborately carved like a fore-edge painting in relief (or should one say a side-edge painting?). No. 129 is similarly carved with figures at a window. This type of fan is thought by Peter Thornton to be Portuguese. He describes one in the Victoria and Albert Museum carved with a garland of flowers round the closed sticks.

No. 8 in the same exhibition (c.1740; lent by Queen Christina of Spain) has remarkably straight guardsticks—perhaps this is a Spanish trait. When folded up they look like a box. No. 34 (c.1760) has very elaborate carved and pierced mother-of-pearl sticks containing painted lacquered cartouches.

Topographical fans are always fun. No. 117 in the Madrid Exhibition was painted by a hack fan painter, probably Spanish, with the market in the Plaza de la Cebada, Madrid; the reverse has the Plaza Mayor (c.1760). The fan was from the collection of the Duke of Alba. The sticks were painted with scallop shells and butterflies, which seems very Spanish in taste. No. 123 shows the Plaza Mayor in 1765. In the Baldwin Collection there was one with a similar view, but a hand-coloured etching. No. 100, from the collection of the Infanta Doña Isabel, is painted with a bullfight in the same Plaza (c.1760), with what appears to be an architect in the foreground.

Lace often appears in *trompe l'œil* fans: see No. 96 in the Madrid Exhibition (collection of Queen Christina of Spain)—a particularly charming mid-eighteenth-century example with yards of lace loosely unfurled right across the leaf. In No. 130 the lace theme also appears in the sticks, and in No. 192 the lace meanders like a path between lunettes.

Very much to the Spanish taste seem *trompe l'œil* fans with irregular-shaped vignettes, often in monochrome. No. 89 is an example; this has the wide-spanning *battoire* sticks favoured by the Spaniards. In No. 118 (c.1740), the irregular-shaped vignettes are echoed in the sticks.

No. 191 in the Madrid Exhibition is a beautiful example of a mica chinoiserie fan. The leaf appears to be mica painted with a bamboo palisade, birds and flowers. Three European portrait miniatures hang in the centre above a chinoiserie couple at tea (c.1760); this fan was from the collection of the Duquesa de Fernán Núñez. No. 158 is a very elaborate fan, decorated with a staircase and arch of straw-work, in a style usually associated with France, but the painting looks very Spanish.

There are some Neo-Classical fans but again, many appear to be of Italian or English origin.

Although it appears that very few fans were produced in Spain in the eighteenth century, they were in fact regarded with very high esteem in the peninsula and the

Below: *The leaf is painted with a game of Blind Man's Buff played by ladies and gentlemen, watched by Harlequin. The sticks are ivory, pierced, and gilt, and decorated with painted fruit. 9ins (23cms).*

Left: English, c.1760. A fan with a leaf painted with an allegory of English trade, with Englishmen, a Dutchman and a Russian. The ivory sticks are painted with flowers and trophies of music, and partly pierced with a trellis. 11ins (28cms), sold at Christie's South Kensington in 1981 for £550 ($907).

Left: Italian, c.1760. A trompe l'oeil fan, the chickenskin leaf painted with prints, a playing card and map scattered across the leaf, signed Michel Artazu in three places; the ivory sticks are silvered. 11ins (28cms).

Spanish nobility ordered a large number of fans from France, England and Italy at that time. I feel that many of the important fan commissions were very precise orders, as the fans from Spain tend to be, on the whole, a little more elaborate and more flashy than those found elsewhere, with far more elaborate sticks. One can only assume that the fanmakers produced a certain number of fans with the Spanish market in mind. Until more research has been published it is impossible to be more definitive. Certainly the high regard for fans in Spain must have led to this safe keeping of such fine examples.

Under Spanish fans, I feel one must include all the French, English and Italian fans designed for the Spanish market. The more one looks at reference books such as the Madrid Exhibition catalogues, the more one can gauge Spanish taste in fans in the eighteenth century. Their taste is very elaborate but it seems catered for almost entirely by imports, except for the topographical fans. Some of the fans painted with bright earth colours and purple may be German; they often have elaborate sticks. Some fans of about 1760, made of net and applied with painted vignettes, are also thought to be German. (Some Germans think they are English, but the English disagree.) It is possible that the fans decorated with elaborate straw-work are German, as are, possibly, fans with articulated sticks, where they are not French. On the other hand, Annaliese Ohm in her 1972 article felt that no German fan industry existed in the eighteenth century. This seems improbable;

Above: Spanish, c.1830. La glorieta paseo de Valencia, *a printed fan, the leaf is a hand-coloured engraving inscribed* Fabrica de Abanicos de Espana y Francia de Fernando Coustellier. *The verso depicts ladies watching babies lying in eggshells in a nest. The sticks are of pierced ivory. 9ins (23cms), sold at Christie's South Kensington in February 1990 for £325 ($580).*

Right: French, c.1760. On this marriage fan the leaf is painted with the scene of a princess arriving to marry a prince. The back of the fan depicts a mansion. The sticks are ivory, carved, pierced, gilt, backed with mother-of-pearl, and painted with portraits of the bridal couple. The front of the fan is original, the verso is 19th century. 11¾ins (30cms).

prints showing fans being sold in German fairs are well known and someone must have provided fans for the proliferation of princely courts. Daniel Chodowicki produced etched fan leaves and Joseph Werner is credited with a number of fans with Baroque decoration.

No. 47 in the Karlsruhe Exhibition is a watercolour fan leaf of 'Winter' by Johan Holzer (1708–40). It is from a set of four, and this artist is also known to have painted another set of four fan leaves. The German fan painters may have been the *Hausmäler* who decorated china from the princely factories. A fan by a Meissen painter, J.C. Hahnemann, is illustrated in R. Haehl's biography of Samuel Hahnemann (Leipzig 1922). The younger Hah-

nemann was a famous doctor, the father of homeopathy. In the exhibition *Fans from the East* (plate 32) there was a strange fan from an Austrian museum painted with a chinoiserie design reminiscent of blanc-de-Chine figures and the work of Pillement, and dating from about 1760. This may even have been made in Vienna: we know that fan leaves were printed there.

So far, the only fans known to have come from Switzerland are a charming group by Johannes Sulzer, 'au Rossignol à Winterthur' (1748–94). They are mainly signed and he also notes that he both painted and mounted them; so it must have been a small factory. There were three in the Karlsruhe Exhibition (No. 61). They are deli-

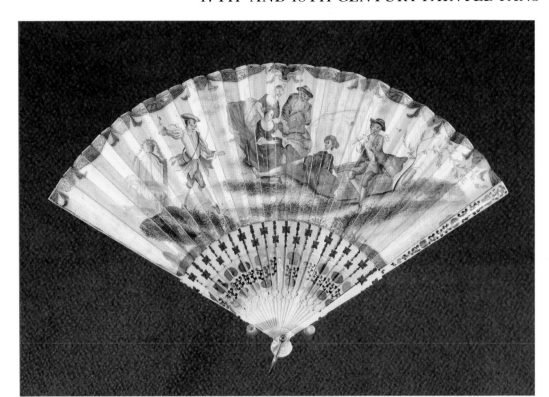

Left: Probably South German, c.1730. The fan is painted in the style of Watteau with musique champêtre. *The sticks are of ivory, decorated with paintings of shells.*

cately painted with small pastoral vignettes and decorated with spy holes in the form of objects such as bird cages. Three were sold at Christie's in 1970; Christie's also sold a similar fan in March 1973, unsigned. There are twelve others in Swiss museums (see the Fan Circle, bulletin No. 11). Christie's also sold an interesting series of Swiss designs for fans in 1973, of figures in fancy dress at a musical evening, some inscribed 'de Flummeren ... St. Gallens ... orn ... 1775'. In January 1980 C.S.K. sold other fans with Swiss views, *c.*1780.

In Sweden, Crown Princess Louisa Ulrike instituted the Order of the Fan in 1744. Britta Hammar has established that fans were certainly made in Sweden in the eighteenth century; the fanmakers registered their marks with local councils and she lists a number of Swedish fanmakers from the 1770s. Kulturen Museum possesses one hallmarked Stockholm 1765 and so is able to tie it to one of four fanmakers who submitted fans that year. Apparently, it compares with the simpler imported articles—Miss Hammar also notes an interesting comparison: Carl Tersmeden in his memoirs writes that he bought two beautiful fans at the Hedemora fair for his sisters at 15 platar each. In the same year he rented two beautifully furnished rooms and attic in a house near the castle in the old town of Gamla Stan for 12 platar a month.

Who were the fan painters? Hardly any signed their work before the last decades of the nineteenth century. Some, of course, were good and will eventually be identified. The brisé painters were often trained as miniaturists, but the paper and silk painters were not, and if they did double as decorative painters, it was in a trade as anonymous as fan painting. A few hands, like that of the damaged fan, C.S.K., 23 October 1979, are distinctive. An English or Anglo-Italian painter of the 1770s who used curiously shaped vignettes, is known from two fans and is almost identifiable, as two similarly shaped drawings by him survive, with inscriptions, dates and signatures, the last unfortunately and tantalizingly just cut off. The painter of C.S.K., Baldwin Collection, 4 May 1978, lot 89, composed his leaf in a different way and was probably the painter of C.S.K., 17 May 1979, lot 56.

With a few Italian painters we know their names, for they signed their fans or designs. Unfortunately, we know not much more. If they are recorded, it is a brief note in Pascoli's *Vita di' Pittori* of 1730 or other such book. At least two Roman firms, 'Camillo Butti' of 1800 and Leonardo Germo, signed their works, as did Sulzer of Winterthur. A small set of fan leaves survive, one apparently signed 'Pietro Fabris'. Four fans in the Chinese manner, said to have been painted by Fr. Castiglione at the court of Quianlong in Peking, were exhibited in New York in 1938.

No French fan painter signed his work, although many people signed fans 'Boucher' and 'Watteau'. Watteau drew Pierrot on a paper fan leaf but did not make a habit of it. Two or three other such drawings probably exist. (See E. Croft-Murray, 'Watteau's design for a fan leaf', *Apollo*, March 1974.)

In England and the Netherlands a bit more is known about fan painters. Thomas Robins the elder painted at least two fans; he was a flower painter and garden designer and among his sketch books, still belonging to his family, is a design for one. Another painter, Joseph Goupy (fl. 1711 on; d.1763), is traditionally said to have done fan leaves, but until a design or a documented fan emerges, we can but guess at his work. He was appointed cabinet painter to Frederick, Prince of Wales, and is most likely to have painted the fan of Cliveden where his patron lived. A third English fan painter was Antonio Poggi, a Corsican,

Above: c.1770. A fan with a paper leaf on which a curiously shaped vignette depicts a classical scene. The reserve is decorated with devices of flowers and peacock feathers on sprigs, the ivory sticks are pierced and the guardsticks are backed with pink tinsel. 11¼ins (29cms).

larity in composition to contemporary illustrators and only a handful of other fans are drawn in a similar way. The reverses of many fans show charming free brush work such as appears on Delft and Lowestoft pottery of the same period. John Smibert, a pupil of Thornhill who practised in Boston, Mass., used to import fans in large quantities; these were probably plain, and so he or a fellow artist could have painted the design wanted. Amateurs abound: the painter of lot 37, C.S.K., 17 May 1979, was probably an amateur who had learnt to draw in the 1770s at Cambridge, where a school of caricaturists had grown up.

As for Dutch fans, Dr. van Eeghen has catalogued the designs of 'Miss' Johanna Alethea Oosterdijk (1736–1813), the daughter of a professor in Utrecht. Among the handful of Dutch fan painters who signed their work is a distinguished classical painter, Frans Xavery, who died in 1768, the last of his line. He signed two excellent fans in Dr. van Eeghen's collection; one painted with the Queen of Sheba is also inscribed 'inv :et', meaning that Xavery had designed the fan, the other, the Altar of Love, is inscribed 'fecit' on a log, and dated 1762, meaning that he painted it but did not design it.

Another Dutch fan painter who signed his work was Johannes Van Dregt (1737–1807). The well-known watercolour painter Dirk Langendijk (1748–1805) is known from one fan, a commission from a print for the silver wedding anniversary of a friend in Rotterdam. A Dutch

who appears to have been a protégé of Reynolds. He published printed fans by Bartolozzi after Kauffmann, and organized an exhibition of fan designs at Christie's in 1782. One of these designs, Angelica Kauffmann's 'Three Graces', reappeared there in 1971; it has been much copied, as it was the basis for one of Bartolozzi's prints.

For the rest of the eighteenth century, fan painters are anonymous figures. Many may have had other artistic employment. Only one painter showed the slightest simi-

Below: An enlarged view of the Dutch theatrical fan seen in the group on page 27.

Above: A selection from a group of fans sold at Christie's South Kensington in September 1989, including: a fan c.1720 showing Diana the Huntress, the verso painted with brightly coloured flowers, the ivory sticks painted, and the guardsticks piqué with silver and clouté with mother-of-pearl, 10ins (25cms), £3960 ($6534); and a fan showing Cupid assaulting a hunting party, the verso with gardeners resting, the mother-of-pearl sticks carved, pierced, and painted with a vignette of Neptune, possibly English, 11ins (28cms), c.1740, £2200 ($3630).

amateur, Lt. d'Art [Lieutenant in the Artillery] van Barenburg, a member of a family of soldiers, signed his *trompe l'œil* fan; he would have learnt drawing as part of his gunnery training. Apart from these, Dutch fan painters are anonymous.

The professionals worked either in the makers' workshops in the cities or at home; one can presume that the better work was done in workshops, often by more than one painter. In Holland, Dr. van Eeghen has pointed out that each workshop had its trademark—e.g. two sections of green fence, or one iron gate and a single green fence: probably Amsterdam—which was incorporated into the design of the vignette on the reverse of the fan. The other painters were probably provincial hacks, possibly journeymen engravers, pottery painters, coach painters and local drawing masters of whom we would know nothing anyway. It might be tempting to attribute examples to distinguished local decorators but the fan trade is too widespread. It is quite possible, for example, that Beilby, the Newcastle glass painter, painted fans for his family, but they could be given to relations elsewhere and so disappear into the wide world. The Meissen painter J.C. Hahnemann painted a fan for his daughter-in-law, but it was his son Samuel's fame as a doctor that led to its survival with its provenance.

In hardly any other European countries are fans signed. The Swedish-born Danish Court painter Carl Gustaf Pilo (1711–93) signed one in 1740 and a Flemish amateur, C. de Stirum, whose family is still well known in Belgium, signed a charming fan in 1792.

Did famous artists paint fans for patrons in the eighteenth century? It is unlikely, as the greatest artists had studios and some assistant could have painted them, and a lesser artist would have been too insistent on his own status to do this: he could have painted a fan for an equal but not for a superior, and if his patron were his equal he would not have painted a fan. But as yet not enough is known to refute the idea completely, and, after all, all famous artists are not rich and famous all their lives. Artists skilled in water- and body-colour painting may have done fans for their relations, and when faced with old age may have turned their brush to a fan. Richard Wilson, the impecunious but greatest landscape painter of the eighteenth century, is known to have designed, if not painted, a pole screen.

If the leaves and fans travelled, however, so did the painters. The Flemish fan painters, with their distinctive colouring of muted purple and tight drawing, probably worked in France and in England. A group of Italians, mainly Venetian or Venetian-inspired, worked in London in the second half of the century. Their training had included 'composition', which English painters had not mastered—had, in fact, not needed to master, as very little decorative painting was done before 1750. Their leaders were G.B. Cipriani and Angelica Kauffmann, but their followers were many and some leaves may also have been painted in Italy especially for the tourist market.

PRINTED FANS

Though printed fans were produced in large

numbers for a mass market, they were

ephemeral and can be as rare today as painted

fans, and just as valuable.

In about the 1720s, the fanmakers, despite their enormous success at the top of the market, found it necessary to expand their trade and produce cheaper fans in bulk to appease an ever-increasing public demand. In order to do so, they began printing the leaves—they could then duplicate their designs and their products could be sold cheaply at fairs on stalls and at booths, thus undercutting the cheap Chinese fans that were swamping the market at the lower end: the dilemma is very well explained in the chapter on chinoiserie in *Fans from the East*.

Printed fans are used to commemorate events such as balloon ascents. They would have been taken home as souvenirs, and obviously sold by the thousand. Being cheap and ephemeral at the time they were often not kept—unlike fans of the highest quality made from pre-

cious materials, their sticks were often only of plain wood. As many of these printed fans are rare, their value often rivals other important fans in price. For instance, whereas a rare mica fan of about 1700 in the Baldwin Collection realized £2,600 ($4,290) at C.S.K. in 1978, and a printed mask fan, *c.*1740, £2,100 ($3,465). Other examples of this charming mask fan exist, together with a nineteenth-century water-colour version in the Metropolitan Museum of Art in New York.

Some fan printers took advantage of the 1734 Copyright Act, which proved to be of great assistance not only to them but also to fan collectors and historians later—just as hallmarking has proved invaluable to collectors of silver. English printed fans after 1734 may have the name of the publisher, often his or her address and the date of publication; sadly, as this was usually at the foot of the

Right: *French, c.1780. The leaf is a hand-coloured etching of a scene set around the Fountain of Love. One Lady, at a window, is being delivered a letter by a dog, while being serenaded by a man with a lute; another lady is being handed a flower by her lover. In the background a boy flies a bird on a string. The reverse carries the words of a song, 'La Fontaine de L'Amour'. The sticks are plain wood. 10³⁄₄ins (27cms).*

Above: *An attractive arrangement of a selection of fans sold at Christie's South Kensington in December 1987. Top left, an English fan, c.1770, painted with a* trompe l'oeil *giving the impression of pink lace, the ivory sticks painted* en suite, *£176 ($290); top right, a French silk-leafed fan, c.1760, painted with a gentleman playing the flute, the reserves embroidered with spangles, and the mother-of-pearl sticks carved and gilt, 11ins (28cms), £176 ($290); middle, a late 19th-century chromolithographic fan-shaped Christmas card, 5ins (13cms), £22 ($36); bottom left, The Margate fan, published on June 10th 1798 by Lewis Wells (Engraver and Fanmaker): it shows seven views of the resort, and the bone sticks are pierced, £550 ($907) (this example slightly damaged); and, bottom right, a fan, c.1905, with a blue satin leaf with silver gauze insertions, embroidered with sequins; the sticks are bone, £198 ($327).*

leaf, it has often been snipped or guillotined during mounting, but as long as one copy survives with the publication line all fans from that plate can be dated. Of course, publishers were not obliged by law to copyright their designs. There was no advantage for the makers of the cheapest leaves, so this is another reason for no publisher's name appearing on a printed fan. (Another amusing point: some dates, names and initials are sometimes disguised in the body of the design, as they occasionally are in painted fans, particularly *trompe l'œil* fans.)

The most important and comprehensive collection of printed fans is that formed by Lady Charlotte Schreiber and presented by her to the British Museum in 1891. It is a fantastic collection of some 734 fans, both mounted and unmounted, of which only about 100 are painted. Most of the printed fans are English—nearly 300; just over 200 are French; the rest are divided more or less as follows: German 23, Italian 18, Spanish 17, 'Dutch' 3, Japanese 1, American 1. As Lady Charlotte was English, it is natural that she should collect mainly English fans, but she was widely travelled and her collection probably provided a good ratio of the sources of printed fans.

In the eighteenth century, England seems to have led the market and exported fans in quantity, sometimes suitably translated versions. Madame de Pompadour in 1760 asks her brother to provide her with printed English fans for presents—these may not have been so pretty, but were cheaper than those that could be obtained elsewhere; see Nancy Mitford, *Madame de Pompadour*.

Certainly some fans have appeared on the market in the last twenty years that are not to be found in the Schreiber Collection, and there are two or three fine private collections of printed fans, and one of fan leaves, but rarely have more than two or three examples of each fan, if that, been seen in the saleroom during the last twenty years. This must prove that printed fans are comparatively rare, except for those lithographic fans of the mid-nineteenth century. (Against this one must remember that the fan printers claimed to print 10,000 of some leaves.) However, one collector has recently made a remarkable find. He bought bundles of unmounted fan leaves from an old print dealer consisting of both Peninsular War propaganda fans printed in Spanish and the-

atrical designs; a high proportion are designs in Lady Charlotte's collection and he feels certain he has tapped one of her sources.

ETCHING

Most of the eighteenth-century English and European printed fans are etchings, the simplest reproductive method of the period, often coloured later by hand. Etching is the method whereby the pattern in the metal plate is made by acid biting away the metal where a protective wax has been scratched away. It is the cheapest method and the most commercial; the artist works straight on to the copper plate. It produces an effect not unlike an indian ink drawing or the finest brushwork of a Chinese painting, and it sometimes requires more than a good magnifying glass to tell the difference.

English eighteenth-century printed fans cover a wide range of subjects. Lionel Cust in his catalogue of Lady Charlotte Schreiber's collection divided them into a number of subsections: Portraits, Historical, The Peninsular War, Classical and Biblical, Fancy, Social, Pastoral, Literature, Painting, etc., Theatrical, Instructive and Amusing. These headings give an immediate idea of the range of subjects covered.

Printed fans often bear titles. These include 'A New Game of Piquet now in play among Different Nations in Europe' in English; another example of this fan, in French, was sold at C.S.K. from the Baldwin Collection. The ten female figures representing France, Spain, Sardinia, Austria, Saxony, Russia, Poland, Britannia, Holland and Prussia are seated round a table with all except the last three taking hands in a game of piquet, Pope Innocent XI declining to take part although his chair is at the ready; on the right stands a commentator and on the extreme right are the Sultan of Turkey and the Shah of Persia. This fan alludes to the intrigues of European diplomacy concerning the affairs of Poland. Madeleine Ginsburg informs me that an unmounted leaf in the Cabinet des Estampes, Bibliothèque Nationale, is dated in another hand 'October 4th, 1733', when the French entered the war of the Polish Succession, before the Partition. About a dozen copies of this fan are known.

Other historical fans in the Schreiber Collection include one depicting the Coronation Banquet of George II, 11 October 1727, with the Champion of England challenging the guests. Sir Robert Strange, the Jacobite, engraved a fan as a memorial to the Jacobite rebellion in 1745, with Prince Charles Edward Stuart in armour attended by Cameron of Lochiel as Mars and Flora Macdonald as Bellona; like many classical scenes, it is rich in allegory.

Right: A colourful selection of French chromolithographic fans from the 1930s: these were valued in 1990 at approximately £80 (c.$130) each. Chromolithography (from the ancient Greek words for 'colour' and 'stone') was a popular printing process in the second half of the 19th century, and, alongside other methods, it allowed the 'mass production' of items such as fans. Originally it was a slow and laborious system, requiring a separate, heavy stone for each colour, but during the course of the 20th century zinc was substituted for stone.

In 1969 Christie's sold a more straightforward version of another incident in the rebellion from the Bompas Collection. Another fan, 'View of the Trial of Warren Hastings, Esq., at Westminster Hall published by Cock & Co., 36 Snow Hill, 22 Sept. 1788' was obviously so successful (as the trial went on so long and was of interest to tourists) that it was translated for foreigners, who looked upon the trial as the pursuit of a corrupt public servant rather than a witch-hunt conducted by a bunch of ignorant and bigoted Whigs. In 1969 Christie's sold one where a Spanish translation has been stuck over the English version. There is one ballooning fan with Messrs. Charles and Robert's Balloon (1783) as well as Biagini's Air Balloon and an accident, although most of the surviving ballooning fans are French.

Other subjects included the temporary buildings erected in Green Park for the splendid firework display of 27 April 1749, to celebrate the Peace of Aix-la-Chapelle. Unfortunately the fan was published on 7 October 1748, and missed all the drama when the grand setpiece caught fire prematurely, and the grand finale with the chief engineer being set about with his walking stick by the German who had designed the display.

As the reign of George III covers most of the period there are, of course, a number of fans alluding to him which were suitable for export both to Hanover and to any country with which Great Britain was allied at the time. These include 'Vive le Roy', engraved by Simpkins and published by T.T. Balster on 19 March 1789; and 'The Royal Family at the Exhibition of the Royal Academy', published by Antonio Poggi on 6 March 1789. It is interesting to note that this fan leaf, engraved by P. Martini after H. Ramberg, is taken from Martini's original plate, the print having been cut down to the shape of a fan. Poggi, a protégé of Reynolds, sold a collection of a hundred of his designs at Christie's on 29 March 1783; he even persuaded the great Bartolozzi to engrave fans for him. Other patriotic fans included George III and Queen Charlotte.

Amongst the earliest English topical fans is a satire on Sir Robert Walpole's hated Excise Bill (1733), which reminds Walpole that it was a similar bill that helped to bring down Cardinal Wolsey. Walpole took the hint and withdrew the bill. There was a sequel, 'The Motion',

Below: Various types of printed fan of the 18th and 19th centuries; an Italian stipple engraving on silk, early 19th century; Loterie d'Amour with wooden sticks and ivory dial set in guardstick c.1773; a chinoiserie etched fan embellished in gold c.1740; an English etched fan of a classical scene decorated with glitter c.1740; a giant fan from the Duchet collection, French 1789; a Spanish etched fan of Christopher Columbus c.1815; an assignat fan; an aquatint fan c.1820; Moses striking the Rock, published by M. Gamble, 1740 – two versions, one embellished with glitter; The Game of Cards by George Bickham, an etched fan, published by Murray in Duke's Court; A Mort de Mr d'Malbrouk, a fan decorated with cut-out paper work c.1770; a lithographic fan with Swiss regional costume, published by J. Kaiser c.1840; and the Queen's Royal fan, published by Thos. Balster 1821, with the arms of the Worshipful Company of Fan Makers.

Above: A hand-coloured etching published by Dyde & Scribe, Pall Mall, c.1780, entitled 'The Art of Divination', listing 36 ideas. The sticks are plain wood. 11ins (28cms).

engraved by H. Franks and published on 6 February 1741.

The marriage of Princess Anne, daughter of George II, with William, Prince of Orange, on 14 March 1734, produced a flow of fans for the English and Dutch markets, many of them highly allegorical, strewn with orange trees. Christie's sold a version engraved by Gyles King and showing the couple attended by Hymen, angels and putti, with in the foreground the verse: 'The Gods in Consult does Agree ... Orange triumphant is again'. The ivory handle closes to form a heart. Madeleine Ginsburg tells me that advertisements by Gamble and Pinchbeck for Royal wedding fans of various types appear in newspapers as early as 1733.

'Maps of the English Channel and of the present seat of War in Germany' were periodically advertised in the 1750s. Fans with maps on the leaves are extremely rare. The Schreiber Collection has a map of England, and a French map of Nicaragua advertising a projected Panama canal. Christie's has also sold a map of North America.

Military matters remember among other subjects the triumphs of Frederick the Great in 'Veni, Vidi, Vici', an English export fan, probably produced in about 1761 as it is an allegory of his triumphs in the Seven Years War; 'The Capture of Portobello', published by F. Chassereau, 22 April 1740; 'The Attack on Cartagena', by Admiral Vernon; and 'Nelson and Victory', 1798—this last also lists '18 New Country Dances for 1799' and was sold by 'Principal Haberdashers in London'.

Amongst portraits there are royal portraits, a fan portraying Sir Thomas White, founder of St. John's College, Oxford, adapted from the *Oxford Almanack* for 1733, and personal fans such as the 'Camperdown Fan, or the

Glorious 11th October 1797', published on 10 March 1798 by Reben, 42 Pall Mall.

Biblical subjects on printed fans again date mainly from the mid-eighteenth century and are therefore etchings. There are comparatively few of them, as these subjects appear to have been reserved for the most expensive market, painted fans. They include 'Moses Striking the Rock', published by M. Gamble, 1740, 'The Birth of Esau and Jacob', and 'St. Paul Preaching at Athens'. These would mainly have been used as church fans.

There is also a small group of classical fans, e.g. Telemachus and Mentor landing on the island of Calypso, and Diana and Nymphs; but there is a larger group, mainly stipple engravings, at the end of the eighteenth century.

The social printed fans are often particularly entertaining. These include 'Mr. Thomas Osbourne's Duck-Hunting', 1754. Osbourne was a publisher and bookseller; on 10 September 1754 he gave a house-warming breakfast with dancing and a band in the marquee and duck-shooting for the gentlemen at his new Hampstead home. The 'New Opera Fan for 1797' is a seating plan with names of the occupants of boxes, published by W. Cock; the King's Theatre fan (1788) has the Prince of Wales and Mrs. Fitzherbert in box 63. There are also fans describing how to play games and to dance the latest dances, such as the Casino fan (1793), published by Sarah Ashton, the Game of Whist fan, and Dance fans for 1792, 1793 and

Right: Another group from the Christie's South Kensington sale of June 1987, top to bottom: French, c.1895. Shown closed, a fan with smoked mother-of-pearl sticks and one guardstick carved with a nymph against a gilt crescent moon, its leaf painted with a misty scene of nymphs by a lily pond, signed Michaels, 14ins (36cms), £715 ($1180). French, c.1895. A double-sided fan, the front painted with an elegant dinner party in a gazebo on a lake, signed G Lasellaz. This artist, Madame G Lasellaz, is known to have worked for E. Kees: five of her fans were exhibited in L'Eventail Miroir de la Belle Epoque in 1985. The verso is painted with the scene of two of the guests leaving the party by boat. The mother-of-pearl sticks are carved with putti, pierced, and gilt. 11ins (28cms), £858 ($1416). French, probably 1885. A canepin-leafed fan which once belonged to Princess Beatrice, daughter of Queen Victoria. It is painted with a couple in a rowing boat, signed Albert d'après L Leloir, 12ins (30cms), £2420 ($3993). English, 1733. A hand-coloured etching entitled 'The New Game of Piquet now in play among Different Nations in Europe', 11¼ins (29cms). £2420 ($3993). Lastly, shown closed, a horn brisé fan of Gothick shape, with pierced guardsticks of gilt metal, 7ins (18cms), c.1810, £104 ($172).

1794 by various publishers. The Royal Connection fan gives the rules of Connection, a card game invented by H.R.H. Princess Elizabeth and the Duchess of York; it was published by Seckars, Scott and Grosky of Friday Street.

Then there are various tourist and souvenir fans for fashionable places: Ranelagh; the Pump Room at Bath, published by G. Speren 1737; the Crescent at Buxton; and the Margate fan, or Guide to the Isle of Thanet, the reverse with a map of Thanet and a directory, published by W. Else (1805). The New Camp Fan of 1794 depicts plans of camps where the militia is being organized in manoeuvres, and advertises itself as being 'sold at all the Fan Shops in London'. Others show 'the Telegraph recently erected above the Admiralty', or the useful 'Almanack 1796 Leap-Year', with an engraving and aquatint by S.J. Neele of 322 Strand, and published by J. Cock and J.P. Crowder of Wood Street.

Some fans were pure entertainment, such as the various Charade fans including 'The New Pick-Nick Charade Fan for the Year 1803', the 'New Moralist or Way to Wealth', or 'The Wheel of Fortune', engraved by J. Fleetwood, 48 Fetter Lane, in mixed media, stipple and etching. 'The Oracle', published by Cock and Crowder & Co. (1800) was a horoscope fan. There is a version in French in the Messel Collection complete with arrow-shaped ivory pointer; another French version, with the verse printed on silk with painted putti and gilt ivory sticks, was sold together with an engraved leaf of instructions by Christie's from the Baldwin Collection in 1978.

Both 'Fanology, or the Ladies Conversation Fan', designed by Charles Francis Badini and published by Robert Clarke, 26 Strand, in 1797, and the 'Ladies' Bill of Fare' were mixed-media fans, line engraving with stipple; the latter had humorous medallions of men as

various kinds of lovers, published by G. Wilson, 14 February 1795, 108 St. Martin's Lane, presumably as a Valentine. Also in mixed media was 'A selection of Beaux' (1795), 'The Map of the Affections', and the 'Lake of Indiferency etc.', dated early eighteenth century (the Schreiber Collection has two versions of the former in French).

Then there is a group of more useful and educational fans, including a fan listing the fares of Hackney coaches, a Botanical Fan, and a History of England since the Conquest, published by J. Cock and J.P. Crowder (1793) and engraved by S.J. Neele, 352 Strand. The Castle Museum, York, has an example of this fan, and there is also a similar fan engraved with the History of France. Particularly useful is the Heraldic Fan, engraved with a series of heraldic symbols with their denominations, published by F. Martin (1792), engraved by Ovenden; this was sold by Wm. Cook, Fanmaker to the Duchess of York, 50 Pall Mall and 55 St. Paul's Church Yard.

There are so many English pastoral and chinoiserie etched fans that I will only single out two less usual ones sold at Christie's South Kensington over the years: a parasol fan, the mount a hand-coloured etching of rural scenes, with wooden handle, 12in (30.5cm) diameter (c.1750); and a fan etched with a musical party with four printed songs and the reverse with a *trompe l'œil* of prints and lace, with sandalwood sticks, engraved by J. Preston (1781). Many others have been sold in recent years. Twenty-four are listed in the Schreiber catalogue and a number were exhibited in the *Fans from the East* Exhibition. They are normally pretty but slight.

These fans form the biggest section of general English printed fans, the overall design being intended for colouring, with all details except the clouds lightly etched in. Although the fans remained in print and were saleable for many years, large stocks do not survive. These fans are traditionally dated to the 1740s and 1750s, and the style and technique of the undated ones confirms this. We cannot, however, date the fans themselves so accurately from this, for a coloured fan leaf in the Schreiber Collection has the inscription in the margin 'Mr. M. pattern 1812'— presumably the date when Mr. M., whoever he may be, ordered a new stock of his cheapest fans. One has then to date the colour of the paints or the shape of the sticks.

Another large group consists of small groups of figures, sometimes theatrical, in etched outline; these are sometimes printed from curiously shaped copper plates. Sometimes different examples exist from the same copper plate with one or more figures removed. It is not known whether these copper plates were made for some other purpose, then cut down and the background details removed, or just made of odd bits of copper. When used on a fan the figures are coloured in and very often two buildings are painted on either side, like a castle or an abbey—usually

Kirkstall. This method saves the printseller the expense of employing a figure painter. The general crudity and weakness of the painting should not put the collector off any of these etched fans, as they were normally issued this crudely coloured. Many mourning fans were made by that method, presumably to be distributed to ladies at funerals.

In the Schreiber Collection is a design for a fan from the *Hibernia Magazine*. This was a folded insert, like the designs for embroidery one finds in the *Ladies' Magazine* (1763 onwards).

A number of etched French fans exist, such as the six designs for fans and a title-page advertisement by Nicholas Loire (1624–79) of Biblical and classical subjects: Isaac and Rebecca, The Finding of Moses, Venus, The Judgment of Paris, Europa, and an Eastern goddess. There are also some by Abraham Bosse (1602–76); Marc Rosenberg illustrates a 1638 fan leaf by Bosse, from the collection of G.J. Rosenberg of Karlsruhe, in his foreword to the catalogue of the 1891 Karlsruhe Exhibition.

Later French etched fans include a pair of mid-eighteenth-century handscreens applied with hand-coloured etchings of the Châteaux of Versailles and Marly, published by Lattre, rue de Jacques, Bordeaux. Dr. van Eeghen has a pair illustrating the story of Jeannot, published by Md. Petit, rue du Petit Pont Notre Dame. There are several fans lampooning 'Malbrouck', the Duke of

Right: French. One of a pair of fans published by Lattre, rue de St Jacques, Bordeaux, consisting of printed handscreens applied with hand-coloured etchings of the Châteaux of Versailles and Marly. (Collection The Hon. C.A. Lennox Boyd.)

Above: Possibly English-made for the French market, c.1785. A hand-coloured etching entitled 'Grammaire Française à l'usage des Rentiers', with wooden sticks. 10ins (25cms).

Marlborough. Others include 'Alzianne ou le Pouvoir de l'Amour' (c.1770), 'Le Triomphe de l'Amour', and Heloise and Abelard. There were also entertaining fans, such as rebus fans in the 1780s where the message is conveyed by reading the images as they sound.

At the end of the eighteenth century quite a number of historical and political fans were produced, these being troubled times: a satirical one, England and America (1778), the Birth of the Dauphin (1781), l'Assemblée des Notables (1787), Louis XVI and Necker, Les Etats Généraux (1789), The Taking of the Bastille (1789), Assignats, Costumes of the Revolution with fifteen figures dressed in the official robes of the revolutionary government, an Allegory of Bonaparte, and the Peace of Amiens. The colours of French printed fans are often more acid and are often cruder than their English contemporaries.

There are also several ballooning fans, including the ascent of MM. Charles and Robert (1783), cabriolet fans and fans with scenes from plays. And, in the third quarter of the eighteenth century, the French produced some giant printed fans.

There are a few Italian etched fans, including seven in the Schreiber Collection. These are mainly classical, and published by Appo. Pagni e Bardi in Via Maggio, Firenze; two of them are engraved by Carlo Lasinio in 1796, one (Cupid and Psyche) is copied from a Bartolozzi

fan leaf or an almost identical Poggi fan leaf, both of which were published on 14 August 1795. Cust in his catalogue presumes it is copied from Bartolozzi, who was a Florentine, but C.A. Lennox-Boyd has pointed out to me that Poggi went to Florence to become a picture dealer. Thus, we have English and Italian fans that are almost identical. They vary only in the smallest details—a Cupid here, a caryatid there. With the three fan leaves in the Schreiber Collection there is a slight difference of size, the Florentine fan being slightly smaller. In about 1800 cheap flag-shaped fans were made mainly in Italy. One example was exhibited, No.47 in *Un Soffio di Vanita*, Padua 1989. Amelia Filippini Sacchetto, President of the Associazione Cultural 'Il Venteglio', published an article on these: 'Ventegli e banderuole in Italia', in 'Il Bolletino dell'Associazione Venteglio', Bologna, Anno 6, No.2, 1988.

Occasionally tourist views are found, such as an early nineteenth-century one sold at Christie's with an oiled paper leaf bearing a hand-coloured etching of St. Peter's, Rome, and ebony sticks.

There are three early Italian fan designs in the Schreiber Collection: 'Battaglia del Re Tessi e del Re Tinta' is an engraving of a hand-screen with a representation of the tournament between the companies of weavers and dyers on the Arno (1619), signed 'Chez N. Bonnart, Jacomo Callot; an early eighteenth-century copy of an etching by Jacques Callot (1592–1635); and a French copy by Nicolas Cochin of an etching by Stefano de la Bella (1610–64), of country dancers. The earliest fan design in the Schreiber Collection is one for a feathered hand-screen with variant

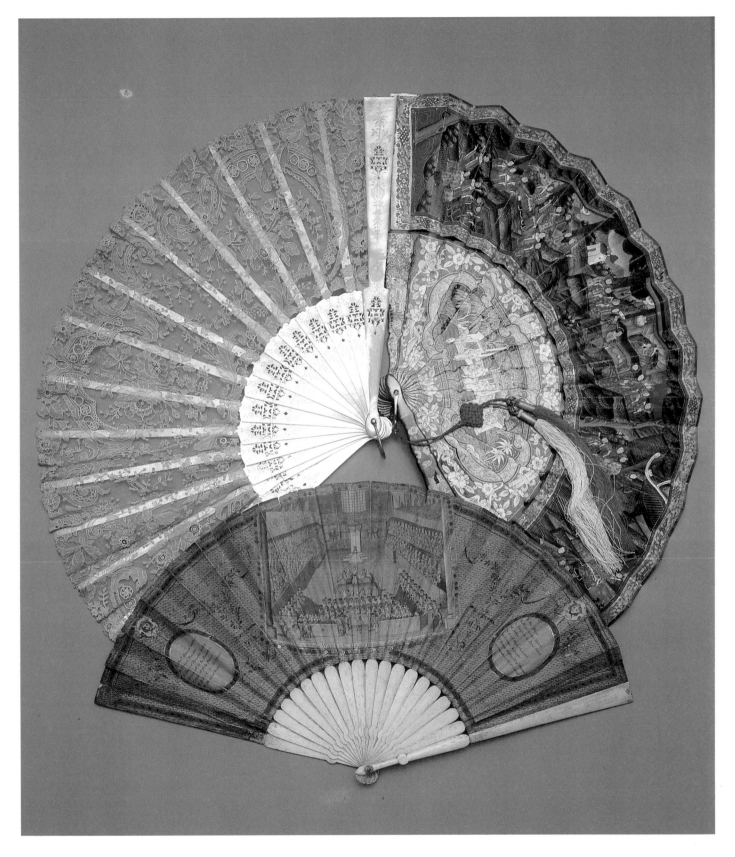

Above: A beautiful arrangement of fans comprising: bottom, a 1788 fan commemorating the trial of Warren Hastings; left, a Brussels lace fan with mother-of-pearl sticks, late 19th century; and, right, a mid-19th-century Cantonese fan.

designs for decorative medallions, by Agostino Carracci (1557–1602); Diana De Grazia suggests it is a headpiece, as it has no handle, and in her *Prints and Related Drawings by the Carracci Family* (Washington, 1979) dates it to 1595.

Spanish etched fans are even fewer in number but include a view of the Plaza Mayor, Madrid, in about 1788.

All but two of the German fans in the Schreiber Collection are etched. An example of a German etched fan was sold in the Baldwin Collection; it depicted Frederick II in Elysium printed on silk. The three so-called Dutch fans in the Schreiber Collection are pastoral etchings, and English, but a small number of stencilled fans, including a ballooning fan in Dr. van Eeghen's collection and a Dogger Bank fan in the Schreiber Collection, are probably Dutch.

Christopher Lennox-Boyd exhibited a Swiss etched fan with three scenes from a Romance, signed J. Kölle Sculpsit, *c.*1790, in Burford in 1987. The style is reminiscent of fans painted by Johannes Sulzer of Winterthur. There was a Swiss industry making souvenir fans with illustrations of cantons and dress in the second half of the nineteenth century; one known maker was J. Kaiser of Lucerne.

In 1990 Christie's s.к. sold an Austrian etched fan, 'Die Sturmende Eroberung von Turkish Sabatz durch Kaiser Joseph II zur erofnung des Feldzug's den 24 April 1788', published by Leonard Schielling, Vienna. There were two other interesting Austrian printed fans in Den Schönen Blicke Zügel, Schloss Bruchsal, 1989 Nos. 69 and 70: a *trompe l'œil* of a medley of objects including fans, published by Waderl Fabrike, Kärntnerstrasse, *c.*1780, the other with an Almanack for 1789 by H. Löschenkohl, Vienna, the verso with a map of Eastern Europe by Johann Hieronymous Löschenkohl (1754–1807).

Above: French, c.1815. The leaf is a hand-coloured etching of a scene from a romance in Tudor period fancy-dress, and the reverse carries a similar scene. The sticks are mother-of-pearl, painted with vignettes and gilt, and the guardsticks are of gilt metal chased with flowers. The fan is set in a double-sided glazed gilt-wood frame.

It is, of course, not that easy to identify the country of origin of etched fans, even if they have inscriptions. An early nineteenth-century fan leaf is copied from two English drawings, or more probably the prints after them by William Williams: 'Before Marriage' and 'After Marriage'; the fan leaf is probably French but it is also inscribed in Spanish as well as French. Another pair of fan leaves are French fans for the Spanish market, each with 32 flowers with their names in Latin and French and the publication lines of 'fabrica de Abanicos de Fdo. Custelier y Compia', with an address in Paris. Another teaser is a fan, 'Grammaire Française a l'usage des Rentiers'; although it is a French 'satirical fan', the Rentier is dressed in English clothes and the etching looks decidedly English.

Nor can the date of the plate be a guide to the date of the fan. The fanmaker Duvelleroy kept a stock of early nineteenth-century plates for his clients to choose from when they wanted old subjects—Dr. van Eeghen has a fan of 1905 with a print of 1845 on it.

AQUATINTS

Towards the end of the eighteenth century, aquatint, a tone process, comes in. Aquatinting is a variation of etching and produces a mass of dots rather like a watercolour wash by the use of a resin that is deposited on top of the wax. Only a few fans using this method survive, as it was rather too complex and elaborate a process.

Examples include a view of Bartholomew Fair in 1721,

published by J.F. Setchel, 23 King Street, *c*.1780, and one of Covent Garden. Six of the Peninsular War fan leaves in the Schreiber Collection are in this medium; they include portraits of Ferdinand VII and of the Duke of Wellington. Most of them are translated into Spanish for shipment out as propaganda, but a number seem to have remained in England unused, in the hoard mentioned earlier in this chapter. 'The Altar of Love', published in 1808 by I. Reed, 133 Pall Mall, is also an aquatint, as is a Conundrum fan (1799).

Although they used aquatints for their finest colour-prints, the French were sparing in using it for fans. In the Schreiber Collection there are three mounted fans of the Revolutionary period and an unmounted leaf of Napoleon. There is also an Austrian aquatint fan of Emperor Francis II. Others elsewhere show The Temple in Paris.

A number of Italian Neo-Classical fans of the 1780s have aquatint leaves, sometimes printed on yellow silk. And amongst the finest of this type of fan are a small number of Spanish fans of the 1820s, commemorating Columbus.

Below: French, 1790/91. A printed fan entitled 'Les Riches du Jour', the leaf a stipple engraving printed in blue with three scenes probably from an opera, the panels headed 'chacun danse a son tour', laissons les danser et fermons les écus', and 'prenes cest le reste de ma fortune'. The verso is printed with a song, and the sticks are wooden. 9ins (23cms), sold by Christie's South Kensington for £385 ($635) in September 1989.

STIPPLE ENGRAVINGS

Stipple was a late eighteenth-century improvement of engraving, a method that produces broken lines of little dots, using a tool rather like a spur. It was a method fashionable under Bartolozzi.

This group of fans, together with the pastoral and theatrical etched fans, are the most common English printed fans. The next group of fans, mainly of classical or fancy subjects, are made from the same copper plates as small fancy prints and are engraved in stipple by Bartolozzi and his followers. On these fans all inscriptions on the plate have been masked out or uninked. The engraving is usually uncoloured and the reserves are painted in with designs. Examples include 'The Theft of Cupid's Bow', engraved by F. Bartolozzi, R.A., from a design by Angelica Kauffmann, R.A.; 'The Power of Love' (1780), engraved by Bartolozzi and published by Poggi; 'Playing Shuttlecock' (*c*.1820); and 'Travellers', drawn by Princess Elizabeth, daughter of King George III, indefatigable etcher and pupil of Benjamin West, and engraved in stipple by H. Thielcke.

There are also portraits of John Milton, Alexander Pope, and Charles James Fox, all unsigned; that of Lt. Col. Sir Banastre Tarleton, the rich Liverpudlian soldier who is best known for his ravaging America with the Light Dragoons (the Tarleton bonnet is named after him), published on silk by I. Cock, Wood Street (1782), was painted by Thomas Stothard and engraved by Wells.

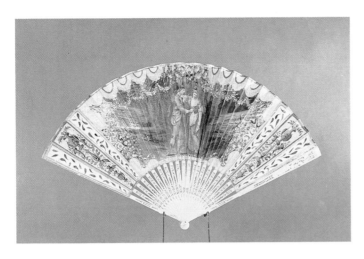

Above: English, c.1790. A fan with the leaf stipple engraved, coloured by hand, and decorated with mother-of-pearl. The ivory sticks are pierced.

Above: English, c.1790. A paper-leafed fan with a stipple engraving printed in colour, depicting a game of Blind Man's Buff. The reserves are painted with swags and peacock feathers, the reverse is painted with flowers, and the ivory sticks are carved and pierced.

Above: English, c.1780. The leaf of this fan is a mezzotint, printed in colours and finished by hand-colouring. The scene depicted is classical, showing the departure of a hero. The ivory sticks are carved, pierced, and gilt. 10½ins (27cms).

Another stipple fan was 'Howard Patriot of the World', *c.*1790, depicting the bust of John Howard the penal reformer (1726?–1790) derived from an etching by Thomas Holloway (1747–1828). Others were the 'Chapel' fan (1796), and 'The United Sisters', commemorating the Act of Union in 1800 of England, Scotland and Ireland, inscribed 'Fair Sister Isles ... blest as free', engraved by George Wilson and published by Ashton and Hadwen, 28 Little Britain, 1 January 1801. The same stipple-engraved vignettes are sometimes printed on silk cut-out and often appear on English fans at this time, sometimes applied to sandalwood brisé fans. The other principal subjects are novels and the classics.

There are several French fans engraved in stipple dating from the end of the eighteenth century. Mirabeau was a popular subject, and early in the nineteenth century there is a group of charming fans, often with stipple-engraved faces printed in colour, and the rest of the fan painted in bright and cheerful colours with romantic and pretty scenes. They are often signed and numbered, and were produced in large quantities. By now the sticks are often very elaborate.

By this date the printed fan had changed its function and country of origin. They are now mainly French or from other Continental countries, no longer plain but normally brightly coloured, and, most important, no longer necessarily cheap. They are often mounted on very elaborate sticks. Publishers include Vve. Garnison.

At the turn of the century, the subjects tended to be commemorative—until about 1815—and then, throughout the Restoration, essentially romantic and pretty fans appeared. The Schreiber Collection contains a single German stipple-engraved fan of King Frederick William III and Queen Louise of Prussia (1798). Dr. van Eeghen has a curious printed fan: its leaf is Italian, with etched outline and stipple features of a Roman scene engraved about 1815, and coloured and mounted about 1845, probably in France; the colours (black and red) are applied without tone, and the shading is in gold.

MEZZOTINTS

The mezzotint is the most laborious engraving process of all, involving covering the copper plate with a mass of scratches and then removing them where the engraver wishes the plate to print light. Mezzotints, although common in England, were very expensive: they cost about £1 ($1.65) when framed in 1760; this was because it is such a very laborious and skilled process.

Mezzotint fans are even rarer than the types already mentioned—only eleven examples and a leaf are known: the leaf (an unmounted *découpé*) may be a Memorial to Queen Mary II, as it is the only known example of a reduction of a memorial print to her. The fans are: 'Departure of a Hero', *c.*1780; 'Soldiers in a Tavern', *c.*1780–90; a French Royalist fan with an oval portrait of Louis XVI inscribed 'Il Reviendera' in sequins, 1793; one of Nelson; three of Lord Rodney—he was notoriously vain and so these were probably private commissions; 'Unveiling a Statue'; a fan with a small allegorical vignette; 'The Night Journey', published and probably also engraved by J. Jehner. Jehner was an obscure but very able engraver and

Above: *A rare mezzotint fan by Bernard Lens. Mezzotint fan are extremely rare – indeed only eleven are known to exist in complete form, and one other as a leaf only. All of these date from the 18th century. The Lens family were miniaturists, and Bernard Lens (1659–1725) produced the fan shown above c.1720.*

painter who worked in London in about 1775, and then in Devon, returning later in 1800 to publish this print, which was originally intended as a transparency. And, finest of all, a fan printed in about 1720 with a mezzotint and with painted reserves, all by Bernard Lens, the second (1659–1725) of the Lens family of miniaturists.

WOODCUTS

Woodcut fans are equally rare. It is a cheap method but probably too coarse for the necessarily small scale of the object and the comparative sophistication of the market. Woodcuts produce a rather crude ink effect, not quite so heavy but with the same effect as the somewhat harsh quality of a child's potato cut or a finger-print.

A French woodcut fan of Necker (*c.*1788) and one other are the only two to have come on the market between 1968 and 1979, and a set of four curious fan designs, possibly Dutch, are the only examples of woodcuts in the Schreiber Collection.

LITHOGRAPHS

Lithography is the most direct process. The plain surface of a prepared stone prints where the crayon is drawn over it. On fans it is easy to identify as there is a black crayon-like under-drawing, and no fan painter would use crayon,

as it would show through the paint. Lithography, although it comes in from the beginning of the nineteenth century, was most used for fans from the 1830s and 1840s onwards; they are usually printed with *fêtes champêtres* in eighteenth-century style and the figures, often wearing seventeenth-century fancy dress, were then hand-coloured. The most elaborate examples have fine mother-of-pearl sticks, the cheaper, bone; the fans of medium quality tend to have ivory sticks. A great number of these pretty fans survive; they tend to realize about £50–£150 ($80–$500) each, depending on quality. Late lithographs were sometimes printed on satin; there is an Italian example in the Schreiber Collection of the Duomo, Florence, by Giovanni Gilardini (No. 219). Printed fans are engraved in the method popular in the country where the print is made. For instance, the French were fond of lithography and produced quite a number of fans by this method, as it is the cheapest way of doing it.

Above: French, c.1860. A hand-coloured lithographic fan showing figures in a landscape dressed in 18th-century costume. The guardsticks are carved with roses, the handle is set with turquoise, and the ivory sticks are pierced and gilt.

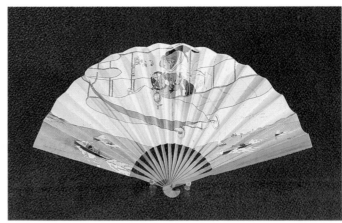

Above: French, c.1918. Many printed fans were produced in the earlier parts of the 20th century as advertisements. This coloured lithographic fan of a lady and a dog in a flying machine was issued by Parfum Pompeia, Parfumerie LT Piver, 10 Boulvd. de Strasbourg, Paris, drawn by Mich and lithographed by Maquet, 10 rue de la Paix. 9¾ins (25cms).

CHROMOLITHOGRAPHS

Chromolithographs are printed from a number of stone 'plates', each in a different colour. They normally have rather a greasy, glassy texture similar to that found on cigar boxes and nineteenth-century chocolate boxes. This process comes in during the 1830s, although it had been invented long before it was first applied to fans. An unusual example with an election scene was sold at C.S.K. in 1978. The prints are often arranged as vignettes within reserves of blue glazed paper.

There is an interesting group of fans advertising international exhibitions from the Great Exhibition of 1851 onwards and including the Paris Exhibition of 1864.

There are fewer chromolithographic and lithographic fans in the 1870s, as there was a fashion for net and cheap lace, although some appeared printed on silk and linen in the 1870s and 1880s, and as imitation late eighteenth-century fans around the 1890s. There is an interesting group of 'art' lithographic fans made in 1904 as a mem-

Below: An enlarged picture of the racing fan illustrated in the group on page 27, showing in greater detail the names of the owners, the names of the horses – including Gladiateur, Blink Bonny, Hermit, Blue Gown, and Beadsman – their sires and dams, and the jockeys – such as Wells and Custance – who rode them to victory. The lithographic process allowed the fans to be produced more cost effectively, and in greater numbers, which made it possible for fans such as this one to be given away or sold very cheaply by advertisers and promoters.

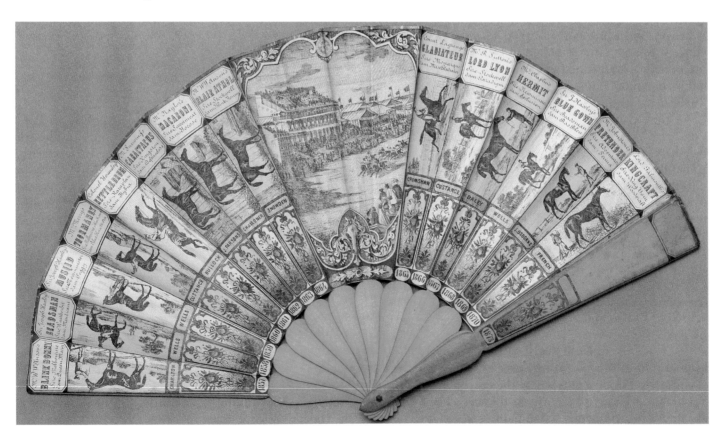

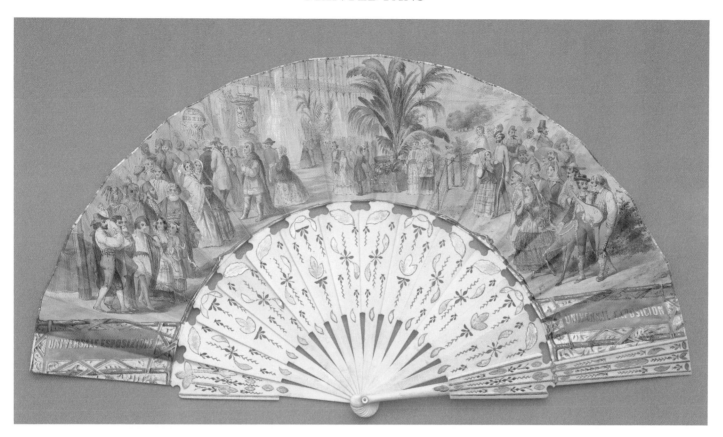

orial to a lithographer, Henry Monnier, by his friends, including Jacques Villon.

In the last two years Christie's have sold two copies of 'Eventail Cycliste', the leaf a chromolithographic map of Paris, published by Atelier Hugo d'Alesi, 5 Place Pigalle, for Leon Pouillot.

The Grafton Fur Company Ltd. sent the Almanak for 1893 to their customers on a fan (it is described as the 20th edition, although this is the only year I have come across). One example sold by Christie's was the one sent to Lady Charlotte Schreiber; the original box was inscribed to her.

Other fans of this period included 'Les Montagnes Russes du Bould des Capucines', advertising Duvelleroy, *c.*1880; one of the Grand Hotel, Monte Carlo, 1880s; and fans advertising the French Railways.

They made a fresh appearance in the early twentieth century as amusing advertising fans—advertising mainly luxuries such as grand hotels (they were probably left at each table for dancers in the evening), restaurants and scents. They were mainly French; certainly the best designed were, such as Charles Barbier for Paquin. These included a series by the publisher Maquet, who employed well-known book illustrators such as Bernard de Monvel. Hotel Knickerbocker published by Duvelleroy in *c.*1910, shows fanciful flying machines over New York. Those fans advertising scents announce that they are appropriately scented, but they seem to have faded over the years, as have, not surprisingly, the late seventeenth-century and early eighteenth-century Italian fans, which are also said to have been scented.

There are also examples of advertising hand-screens

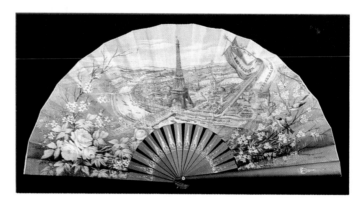

and cockade fans. A number of cheaper advertising fans were made in Japan, often with woodcut decorations; one appeared in a sale at c.s.k. in 1979 advertising Dutch cocoa. Advertising fans can still occasionally be encountered on airlines, chiefly those from the Far East. And the 1978 invitation to the Royal College of Art's Fashion Show was a chromolithographic paper brisé fan.

It is strange how many of these late printed fans survive, often in mint condition, but they were probably kept to remind the recipient of a special evening—a romantic dancing partner, or a delicious dinner—and, of course, folded fans take up very little space.

NINETEENTH- AND TWENTIETH-CENTURY PAINTED FANS

Fan-making in the early 1800s was big

business – by the late nineteenth century it had

become high art. The flow of design ideas led to

increasingly international styles.

In the nineteenth century fans become more mass-produced and, on the whole, less fine in quality, but they are very decorative. Spain joined France in becoming a leading producer. Painted fans are no longer necessarily the more important fans. The painted scene is often relegated to the back of the fan and the front bears a lithograph such as one sold at c.s.k. from the Baldwin Collection, a fan depicting the Surrender at Baylen; the reverse is painted with the Maid of Saragossa, one of the heroines of the siege. General Pierre Dupont with 18,000 men surrendered on 23 July 1808 to General Castanos at Baylen. Among the results of this severe reverse to the French, the siege of Saragossa, an unfortified city and the capital of Aragon, was temporarily raised. It was renewed at the end of December, the defences were breached on 27 January, although the city did not fall until 20 February after an heroic resistance. Maria Agustín, the Maid of Saragossa, is mentioned in Byron's *Childe Harold*, Canto 1. This, like many other early to mid-nineteenth-century fans, is painted very much in eighteenth-century style, notably in the arrangement of three vignettes on the leaf.

In the 1840s to 1860s the figures are often dressed in eighteenth- or even seventeenth-century clothes and are enjoying *fêtes champêtres* and other pastoral and courtly pleasures—the same pastoral pleasures that they enjoyed on eighteenth-century fans.

Most of the early nineteenth-century fans are small

Above: *English, c.1820. The leaf is painted in the style of Stothard or Corbauld, featuring a scene from a novel.*

ivory or horn brisé fans, some English, others possibly made in Dieppe, pierced and painted with garlands of flowers. Sometimes, however, they are more elaborate and are painted with charming vignettes such as double-image fans, or the unusual fan, No. 127 in the Madrid Exhibition in 1920, which is rather an odd mixture—a pierced horn brisé fan of Gothic shape painted with a chinoiserie frieze.

Fans of this type date from the first quarter of the nineteenth century; there were, however, similar fans

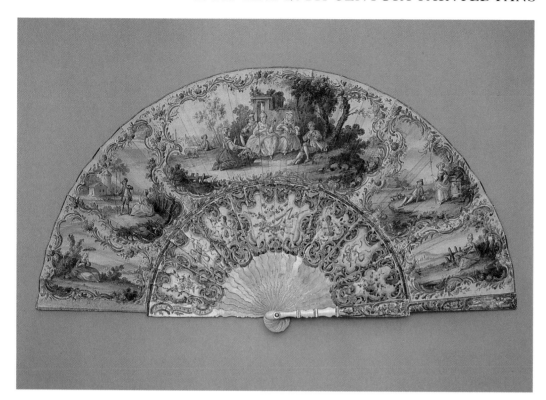

Left: French, mid-19th century. This is a pastiche of an 18th-century fan. Its leaf is painted, the recto with ladies and gentlemen in a park, and the two vignettes show a fisherman, and a lady with a bird on a string. The verso depicts a girl offering flowers to a group of elegant figures, and also carries four smaller vignettes. The ivory sticks are carved, pierced, and painted with trophies of love. 11ins (28cms), sold by Christie's South Kensington for £352 ($580) in February 1990.

Above: A 19th-century pastiche of an 18th-century fan design: the mother-of-pearl sticks date from c.1760. The leaf is painted with an elegant flautist and his audience, and with personifications of the Arts. The reserves are painted with shells, vines, fruit, feathers, and flowers. 11¾ins (30cms).

made late in the century. From the first few years of the century, there survive some fine English brisé fans, extremely well painted, probably by miniaturists, particularly those who specialized in illustrations, such as William Marshall Craig. One was painted commemorating the Act of Union of England and Ireland, 1800. In Spain and Italy during the same period (first quarter of the nineteenth century), there are examples of small fans, often with fine leather or chickenskin leaves, rather fine sticks, often of mother-of-pearl, and gilt metal guardsticks set with semi-precious stones. A fine example of this type of fan was No. 152 in the Madrid Exhibition (from the collection of the Duquesa de Talavera); it is painted with a palace in a park with temples—la Alameda de Osuna—within red drapes. The reverse is painted with the initials of Doña Maria Josefa Pimentel, Condesa-Duquesa de Benavente, wife of the Duke of Osuna who built the Alameda and was called 'El Capricho'.

In the 1820s and 1830s fans grew longer again and the Spanish ones, particularly, resembled eighteenth-century ones—even the mother-of-pearl sticks are carved and pierced in eighteenth-century style. Such fans were probably produced only in France and Spain. No. 230 in the Madrid Exhibition is painted with the wedding procession of Ferdinand VII and Doña Maria Cristina. The leaf is a full 180° and the procession winds right round the semicircle as it might in an oriental fan, but an uncommon occurrence in European fans.

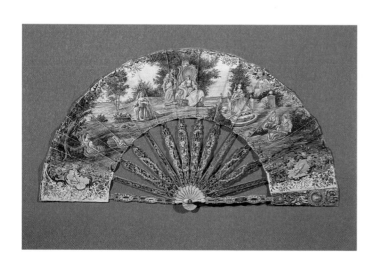

Left: The leaf is mid-19th century, the sticks c.1770. The painting is of classical figures in a park with a rococo fountain. The ivory sticks are carved and pierced with putti, and the guardsticks are set with vignettes, one of the Altar of Love. 10¼ins (26cms), this was sold by Christie's South Kensington for £330 ($544) in May 1989.

Above: *English, c.1855. A charming fan which belonged to Queen Victoria. The painting on the leaf is signed by T Boultry, and shows characters dancing in Spanish costume. The sticks are carved with the royal arms and full-length portraits of Queen Victoria and Prince Albert. Sold by Christie's South Kensington for £3960 ($6534).*

Above: *A fine fan showing a lady, perhaps an extremely elegant shepherdess, in a pastoral setting, surrounded by putti, being watched by a man hidden in bushes. The artist, A Solde, has signed the leaf by showing his name carved onto a stone plinth. This was sold by Christie's South Kensington in March 1983 for £1100 ($1815).*

Now, more amateur artists are embellishing their own fans, although they do not do so in quantity until the second half of the century.

Pairs of hand-screens were fashionable in the second quarter of the nineteenth century. Most are of papier mâché painted with landscapes and lacquered; some are of pressed card. Many are of English origin.

Until the last few decades of the last century, when fans began to be signed, the anonymity of fan painters remains. In England, however, some names appear. Richard Doyle, the caricaturist and illustrator, painted a fan which is now with the collection of H.R.H Princess Margaret.

By the mid- and later nineteenth century, most fans are hand-coloured lithographs. Only the very finest fans, at least the most expensive being some made by Alexandre, and some of the weakest fans, often by amateurs, are now painted. From this point onwards, it is common practice

for painted fans to be signed. A quantity of Spanish and French fan painters flood the scene. One fan sold at C.S.K. was even signed by three artists, as there were painted vignettes on the sticks. In Paris in the 1870s, several distinguished artists painted fans, such as Olivier de Penne.

In the 1860s and 1870s fans were often painted on satin. For amateurs, Duvelleroy and the other fanmakers sold plain satin fans and ivory and wooden brisé fans to be painted with flowers and birds. Apparently, a dog's head or a favourite racehorse were popular subjects in 1886, according to the *Englishwoman's Domestic Magazine*. Some autograph fans fall into this type; they are normally painted on brisé fans, each stick having a small vignette and an autograph or inscription.

Although the mainstream of High Victorian illustration—'the illustrators of the Sixties', in James Gleeson White's phrase—was in subject and treatment unsuitable

Above: A mixed group of fans from a Christie's South Kensington sale in May 1990, including: a palmette-shaped fan by Duvelleroy, Paris, painted with a white cat, the verso depicting the back view of the cat's head. 10ins (25cms), c.1905; and a shaped wooden brisé fan, also by Duvelleroy, painted in gouache with the head of a Pekinese dog, the verso showing the back of the dog's head, the guardstick with an elegant whip, 5¹/₃ins (13.5cms), c.1910.

for fan painters, there were painters whose work could have been adapted for the fan. But we must wait until another fan can be identified. In an attempt to improve the quality of fan painting, the Fan Makers Company launched a series of exhibitions and competitions in the 1870s. Unfortunately these were not a success.

On the Continent, illustrators mainly used lithography. Unfortunately, it was rarely coloured and so the school of illustrators was, with the exception of Romantic subject painters like the Deveria brothers, far removed from fan painters, and had no such similar trade to turn to or be trained in, and so we cannot even hazard a guess under whose influence the fan painters worked.

The 1880s and 1890s saw a revival of more interesting and artistic fans (F. Houghton and his daughter Rosie were good English fan painters of this period.) There are some superb and very amusing, often eccentric, designs

exhibited in the Karlsruhe Exhibition (1891), including nude ladies on polar bear skins, and giraffes knocking putti out of palm trees. Another very amusing fan was sold at c.s.k. by S. Drinot for Duvelleroy of a policeman and children stealing apples: a double-sided fan, it is said to have been designed by Richard Doyle for Louise, Lady Hillingdon. Another, of a circus, signed J. Donzel, had a clown on the guardsticks.

Cheaper or amateurish fans could be rather ridiculous, such as the one sold at Christie's of a lady promenading her poodle along a twig.

Above: English, c.1880. 'Hunting Family', a comic fan painted with four scenes of the Huntbatch Family out hunting after a drawing by Randolph Caldecott. The fan has pierced bone sticks, and was sold by Christie's South Kensington in May 1989 for £605 ($998).

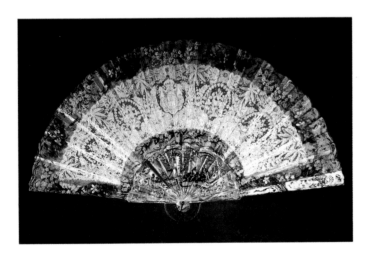

Above: c.1880. The leaf is of Brussels lace, worked with a coat of arms within a border of gauze which is painted with flowers. The coat of arms is believed to be that of Princess Annie de Lusignan. The mother-of-pearl sticks are carved, pierced, silvered, and gilt. 14ins (36cms).

Right: The leaf of the fan above is of Brussels point de gaze worked with lilac, roses, and ferns: the shaped silk vignette is painted with a lady standing by a lily pond with two putti, and is signed E Guerard. It has mother-of-pearl sticks carved and pierced with putti. 11ins (28cms), 1893, sold by Christie's South Kensington in September 1987 for £1100 ($1815). The fan below in the case inscribed 'Queen Victoria's Fan' was seen in detail on page 64: it was presented by Queen Victoria to Lady Charlotte Locker, then passed down in her family until eventually auctioned at Christie's South Kensington on 22 September 1987 for £3960 ($6534).

Above: French, 1878. The leaf is painted with a bunch of pansies and violets tied with blue ribbon and the royal arms of Sweden and Denmark, signed Mathilde, 1878, the reverse painted with the royal arms of Sweden in a garland of violets. The fan has tortoiseshell sticks, one guardstick set with the initial L, crowned, 11½ins (29cms).

Another type of fan of this period was often painted on gauze with lace insertions specially worked to the correct shape, sometimes forming objects such as gondolas, and in the case of a pretty fan sold at C.S.K. from Lady Charlotte Schreiber's collection, a gazebo. They can sometimes be viewed from both sides. Billotey, who is known as a painter on glass, painted some very fine fans on gauze of flower and insect subjects in the 1890s. The cheaper versions of this type of fan were less well thought out, just painted with sprays of flowers or butterflies interspersed with insertions of fragments of lace flounces.

Occasionally, one comes across rare and interesting fans, painted and used by aesthetes. One mount was painted with three portrait heads by a follower of Sir Edward Burne-Jones. The fan came from Old Battersea House, the house of Mrs. Stirling, patron and collector of her brother-in-law William de Morgan's pottery, and was probably painted by her great-uncle J.R. Spencer-Stan-

Right: German, c.1890. The painting by C Auront on the gauze leaf of this fan shows Prince Herman of Saxe-Weimar winning the Brown Derby in Germany. It was formerly in the collection of Princess Gerta of Saxe-Weimar. 17ins (43cms).

Above: English, late 19th century. A particularly interesting fan painted by a follower of Sir Edward Burne-Jones, probably J.R. Spencer-Stanhope. The kid mount is painted with three portrait heads, a serpent, oranges, and pomegranates. 10ins (25cms).

Above: Late 19th century. The black lace leaf is set with a shaped vignette painted with lovers and putti by a gazebo. The sticks are stained mother-of-pearl. 14½ins (37cms). The fan belonged to Lady Charlotte Schreiber.

hope, a friend and pupil of Burne-Jones. In May 1989 C.S.K. sold a rare Arts and Crafts fan, the dark blue leaf painted with pairs of dragonflies (c.1890).

During this period, until the First World War, almost for the first time, artists of high quality, as opposed to fan painters, are known to have designed and even decorated fans. Charles Conder designed many such fans. His fan leaves were mainly painted on silk with pretty pastoral scenes: they were considered too good and too delicate to be mounted. Amongst other examples, there is one in the Ashmolean Museum, Oxford, and they can be found passing through the London salerooms regularly.

Whistler and the interest in Japan from the 1870s onwards had clearly re-awakened the interest in fans; although fans were no longer a fashionable accessory, the fan shape was fashionable and still used. Edgar Degas decorated fans with crayon drawings of dancers and there are examples by Van Gogh, but fan designs by artists of this calibre tend to appear only in picture sales and important art collections. In the twentieth century, the German Expressionist Oskar Kokoschka painted fans; an exhibition of them was held in 1969. There were also less famous artists like George Sheringham, who held an exhibition of fans and panels painted on silk at the Ryder Gallery in 1911. Others include John Dickson Innes, William George Robb, and Sidney Bagoff de la Bere.

In France, decorators such as Barbier were designing fans: there is a fine lithographic example of his work. Erté designed some unusual fans, but one does not know how many of his fanciful designs were actually carried out. The

Right: c.1890. The chickenskin leaf is painted with birdcatchers and other figures in 18th-century costume set in a park. The verso bears five putti bearing garlands of flowers and blue ribbons. The sticks are mother-of-pearl, painted with vignettes of country scenes. 11ins (28cms). This fan was sold at Christie's South Kensington in December 1988 for £1100 ($1815).

Above: This mixed group illustrates the very wide range of styles produced in different periods and regions. There are three 19th-century fans among them. The bottom fan in the grouping on the left is Japanese, c.1880, decorated with hand-coloured wood-cuts of autumn and winter scenes, 15ins (38cms). The fan just above it, on the right, was produced in roughly the same period (c.1890) in Europe, probably by Duvelleroy, and depicts a pastoral scene, with a man holding Cupid and his bow over a hedge. 9½ins (24cms). The third is Spanish, again late 19th-century, painted after the style of Goya, inscribed with the motto 'Lo que ha de ser, no peude faltar'.

Dutch Art Nouveau painter Willem van Konijnenburg (b. 1868) painted a wood brisé fan for a friend in 1890.

Feathers, lace, sequins and lithography were playing such a large part in the decoration of fans in the late nineteenth and early twentieth centuries that, apart from the mass-produced cheap fans with small scenes and some bullfights hashed out by Spanish fan painters, very few painted fans were produced at that time. There were a few special commissions from important serious collectors, such as Queen Mary, the rich Haut-Bohemians commissioning fans from artists, and the French designing way-out fans to tone in with way-out clothes—such as Barbier for Paquin, or Erté for himself.

Right: Details from two fans of the second half of the 19th century: left is a silk leaf depicting purple and yellow iris and insects, painted by Billotey c.1880; and right is a Brussels lace fan with large shaped silk insertions, painted with a little girl arranging a garland of flowers in her hair and attended by putti, signed by G Nietes, c.1865. Below is an earlier fan, c.1700.

ORIENTAL AND ETHNOGRAPHIC FANS

Beautifully carved fans imported from China

had greatly influenced European fan design;

however there are centuries-old traditions of

fan-making in the East, Africa and the Americas.

Apart from different types of fan, in China from the late seventeenth century onwards two distinct styles were produced: those for internal consumption and those for export to Europe via Canton and India. These, in turn, can be subdivided.

Before that date one is dealing with innumerable types of fans, all for home use. The collector today is unlikely to be able to add any examples to his collection, although there are some fine early Chinese fans in museums, mainly unmounted or dismounted painted fan leaves.

CHINESE AND JAPANESE FANS

The history of Chinese fan painting is the history of Chinese painting. The student of Chinese painting is not helped by the great reverence for past masters and past styles that has existed in all Chinese art. The scale and medium of painting have remained constant: brush and ink outline with faint colour washes laid on paper or silk. The subject is normally landscape—usually sublime and misty. Distance is indicated by emptiness, often subtle. This, with the diagonal composition that was the norm until late in the Yuan period, means that fan shapes, both rigid and folding, are most suitable for paintings. For many centuries there were schools of 'literati painters', rich officials who dabbled in art and exchanged paintings, poems and calligraphy. All this led to fan decorations becoming a major part of painting and calligraphy.

In Japan the reverence for Chinese forms existed from the beginning and, even if they departed from the canons of Chinese taste, they did not often stray far. They painted

Right: Japanese, late 19th-century. Here the reverse is shown to illustrate the delicacy of the painting technique used to create the butterfly and wild flowers. The front of the fan depicts a pedlar, with a river and mountains in the distance. 11ins (28cms).

Above: Amid this attractive group are four Oriental fans of different periods and types. Bottom left is a Japanese fan made c.1880 showing a lady in a rickshaw and other figures in a landscape, the guardsticks decorated with shubayama-work, 13ins (33cms). Just above it is a 19th-century Japanese war fan, the leaf painted with symbols of the sun and the moon against a gold and black ground, with wooden sticks and iron guardsticks simulating bamboo, 11½ins (29cms). The dark-coloured tortoiseshell brisé fan above and to the right of the war fan is Cantonese, c.1830, 7ins (18cms). The light fan in the middle of the group is Japanese, c.1875: it is ivory brisé, lacquered in gold and silver and decorated with a painting of cranes in a mountainous landscape, 9ins (23cms). These four fans were sold at Christie's South Kensington in February 1988 for £440 ($726), £1650 ($2722), £440 ($726), and £1320 ($2178), respectively.

flowers and fish more often than the Chinese and they used bright colours when they painted people—which they did often—and they painted on gold leaf. None of these differences interfered with their fan painting.

The lack of shadows and perspective made fan composition easier, but when in the late eighteenth century these Western ideas crept into Japan, it was the popular painters of Edo who used them; these were the artists of the block prints so famous in the West, who worked for the ordinary tradespeople of the capital. They designed prints to go on fans and in these they added perspective. A few of these Japanese printed fans survive, and a dismounted example from the Victoria and Albert Museum by Utagawa Sadahide (1807–73), depicting a seaweed gatherer, was exhibited in *Fans from the East*.

Many of the Chinese and Japanese fans for 'home' use are of the painted variety. They often survive only as unmounted or dismounted fan leaves, for both folding fans and hand-screens, because it was the painting which was valued. Indeed, the sticks of Chinese fans for use in the East are normally very plain, often bamboo, as opposed to the elaborately carved and lacquered sticks produced for the European market.

There were of course a number of other types of fan made in both China and Japan at the time; these were mainly ethnic and chiefly hand-screens. A Qianlong (1736–95) hand-screen of jade with gilt decoration from the Fitzwilliam Museum was illustrated (plate 4) in *Fans from the East*. A flywhisk in the Horniman Museum has an ivory handle decorated with strands of coloured silks with an overlay of hair. In the Victoria and Albert Museum is a pair of fine early nineteenth-century hand-screens of tortoiseshell and bamboo appliqué with embroidery.

There also exist several nineteenth-century folding fans with maps, more similar to European fans. The Horniman Museum contains a late nineteenth-century hand-screen

Right: Chinese. The leaf is painted with a biblical scene, after a European design, the figures in a landscape near a river, watched by lions. The sticks are mother-of-pearl, carved, pierced and gilt, and inscribed SM 1786. The case in which it is seen here is 20th-century.

Above: Cantonese, early 19th-century. The narrow leaf is painted with figures in a landscape, and the ivory sticks are carved, pierced, and painted in bright colours with more figures, but in a much cruder style. The reverse shows women picking flowers on the leaf, and on the sticks, a similar design to the front. 13ins (33cms).

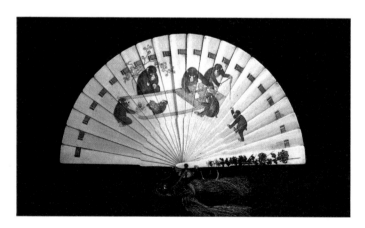

Above: Japanese, late 19th-century. An ivory brisé fan with hairwork carving of saru examining a kanemono, and one holding a peach plucked from the tree seen behind. The guardsticks are decorated with gilt takamakie. 11ins (28cms).

of betel palm decorated in poker-work, and a hand-screen of palm leaf with a bamboo handle decorated with tortoiseshell. Similar but simpler fans of this type are still made in China, and at the time of writing were being used for window-dressing in Peter Robinson in London and in a smart dress shop in Amsterdam!

In Japan, the *Suehiro ogi* or 'wide-ended' fan first appeared in the fifteenth century, but was still used in the nineteenth century. There were also heavy curved wooden hand-screens, including those used by umpires at wrestling matches, folding fans, and hand-screens with iron handles used by commanders for giving orders in battle.

The earliest Chinese export fans to survive are a small series of ivory brisé fans lacquered in red and gold with scenes depicting both European and Chinese figures, the reserves with flowers and beasts, and the borders pierced. The example in the Victoria and Albert Museum (2256'т6) was exhibited in *Fans from the East*. The Europeans depicted in these fans look rather strange, as the Chinese artists were not able to understand how their clothes worked. This we shall see again and again in the Chinese attempt to copy European fans or draw European figures—to please and capture the European market—as it is usually in the clothes and in the faces that they fail.

The next group of fans to survive are painted fans of about the 1780s copying English eighteenth-century fans with Biblical and classical scenes, such as the fan in this author's collection. This has mother-of-pearl sticks and is carved with initials and a date, and must have been a special commission.

From the same period more fans survive painted with Chinese figures in a landscape against an off-white/brownish ground, mounted on plain pierced ivory sticks with carved guardsticks. The leaves of these fans are sometimes *découpé*, possibly done later on arrival in England. These fans rather resemble the English etched chinoiserie fans of the 1740s. In 1979 c.s.k. sold an example which belonged to Lady Charlotte Schreiber. A fine example sold by c.s.k. from the Baldwin Collection is now in the Victoria and Albert Museum. This was painted with Chinese figures with applied silk clothes in an interior; the windows are mica and net insertions, and the reverse of the fan shows the exterior of the house. The sticks are lacquered and the guardsticks are ivory carved with a tree and its roots.

Another charming group of Chinese late eighteenth-century fans is painted with views of the European merchants' warehouses or hongs at Canton. An example in the Victoria and Albert Museum was illustrated (plate 6) in *Fans from the East*, and a similar fan in the Mottahedeh Collection in A. Du Boulay, *Chinese Porcelain* (figs. 121 and 122). The reverse of such fans is painted with flowers and fruit.

It was this style of painting—brightly coloured flowers, fruit, birds and figures—which was to enchant Europeans, particularly the English, for the next hundred years. The colours are clear and bright and the draftsmanship is very precise, like that of a miniature artist. From about the turn of the century to about 1820 one finds fans painted with these pretty birds and flowers, and with pierced ivory sticks.

From about the 1790s onwards survive a large number of pierced ivory brisé fans, at first long and narrow but getting shorter and wider by the 1810s–1820s. A thicker and clumsier fan appears after that. The later ones are often pierced with decoration against a ribbed ground rather than carved, and often have a central plaque carved with initials, again special commissions. The later they are, the more Chinese figures and scenes are crowded on to the fan. There are also rarer examples of parasol or cockade ivory brisé fans with long handles. Towards the end of this period one also finds mother-of-pearl brisé fans, but these are much rarer. There are also brisé fans of cloisonné enamel, of sandalwood, and of tortoiseshell, but these, again, are rarer.

At the same time black and gold lacquer brisé fans were made. They were often similarly decorated to the ivory fans but painted, not carved or pierced, with a central shield with initials and, characteristically, an over-all small motif.

From about 1820 one occasionally finds fans which resemble a brisé fan with a narrow mount attached so that the sticks take up two-thirds of the length of the fan.

From about the 1820s we come to what, with the ivory brisé fans, are the most common of Chinese export fans:

Below: A Cantonese carved ivory brisé fan, c.1820. The elaborate devices include lords travelling with their servants; hunters; pagodas; real and mythical animals; and a central panel containing a monogram.

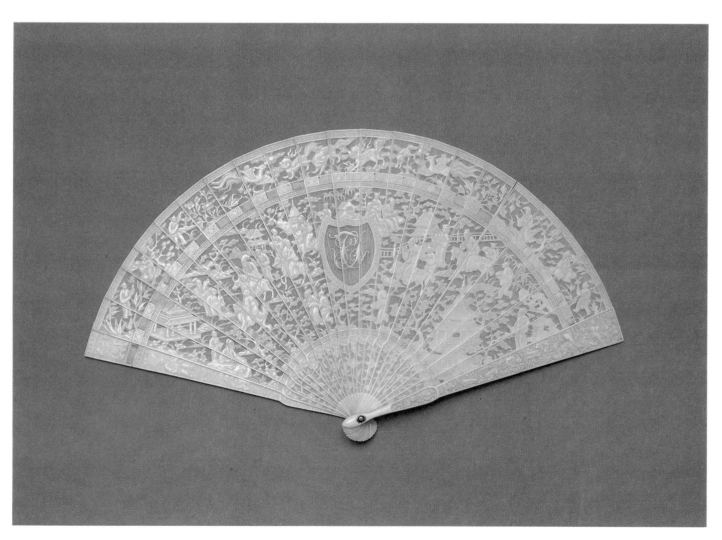

the fans applied with figures with silk clothes and ivory faces which have been known as 'Mandarin' fans. These fans were painted with scenes from Chinese life and as the century wore on they became more and more peopled. The amount of detail and number of applied ivory faces was quite a feat. The reverses, if not similarly decorated, were painted with flowers, birds, fruit and insects, sometimes on a silvered ground. The sticks were usually of pierced and carved ivory, though they are also to be found with sticks of stained ivory, bone, cloisonné enamel, lacquered wood, filigree and sandalwood. These Mandarin fans were also made as cabriolet fans on occasion, and as telescopic fans.

During the 1820s, an interesting series of fans was produced in Macao for the Portuguese market. These had elaborate allegorical painted paper mounts depicting the Portuguese Royal family. The reverses were painted with brightly coloured European flowers. They were rather long – 12½in (31cm) – as was becoming fashionable, and had gilt filigree sticks.

Also about 1820, the Chinese exported charming feather fans made of white goose feathers painted in bright colours with birds and flowers in the same style as the painted paper fans; they had ivory or bone sticks.

In the 1880s a number of fans with asymmetrical sticks were produced.

ORIENTAL, OCEANIC, AND AFRICAN FANS
Between Japan and China lies Korea which was a hierarchical Buddhist state and, not unnaturally, had fans. The Pitt-Rivers Museum in Oxford has two interesting examples given by the Royal Botanical Gardens, Kew, in 1888. They are hand-screens made of rays of split bamboo covered with painted broussonetia paper: one has a free swirling motif in orange, green and yellow (the Korean

Above: A colourful group of fans from a sale at Christie's South Kensington in February 1990. Seen top left is an Orientally influenced French early 19th-century fan, the leaf painted with chinoiserie figures in a garden, their faces of ivory, their clothes of silk: a screen at the centre of the back contains a mica spy-hole. The leaf is painted with garlands of gold flowers with mother-of-pearl petals, and the sticks are lacquered in black and gold, 12ins (30cm). Bottom centre is an early 19th-century Chinese fan, the leaf painted with a vignette of a shepherd and shepherdess in European dress, the reserves and verso with sprays of flowers, the filigree sticks gilt and cloisonne with blue and green enamel, 9ins (23cms).

national colours)—very Art Deco; the other has black Art Nouveau flowers against a gold ground. Two others in the Horniman Museum are painted in the Korean national colours, and bear the national emblem—one is over three feet high. All have plain wooden handles.

The Koreans also had folding fans. The Botanical Gardens presented one such fan to the Pitt-Rivers Museum in 1880, a plain fan of broussonetia paper with bamboo sticks; the only difference from a normal European fan being that its thick handle curves slightly.

Farther south, in Malaysia and Indonesia, there is a distinctive type of fan, mainly pear- or leaf-shaped handscreens of buffalo hide painted in gold with dancing figures and with horn sticks. Although solid they are translucent, and rather decorative. Folding examples are much rarer but do exist, mainly from Sumatra.

From the island of Java, to the east of Sumatra, come multiple cockade fans of palm leaf: an example is in the Horniman Museum. Another such fan comes from the island of Praslin in the Seychelles and is illustrated in *Fans from the East* (plate 46); this was identified as being made from the coco-de-mer that grows only in these islands.

In the South Sea islands fans probably achieved their greatest social importance, and some of the largest fans were made. They are almost all of woven strips of palm leaf, the shapes varying from archipelago to archipelago; the higher the status of the holder, the grander the handle.

Above: A Cantonese fan, c.1840. The ivory sticks are carved with butterflies and flowers and decorated with feathers. The leaf is painted with military scenes: the warriors have ivory faces and silk clothes. The fan was sold at Christie's South Kensington in May 1989 for £770 ($1270).

Above: Cantonese, probably mid-19th-century. This large fan is fashioned in silk over paper. The leaf is painted with figures which have ivory faces and silk clothes. It has wooden lacquered palmette sticks. 14¾ins (37cms).

In the Pitt-Rivers Museum there is a collection of the humbler fans, including a miniature fan decorated with seeds and beads from Melanesia, purchased from the Hooper Collection in 1931. In the Marshall Islands (in Micronesia) the fans are diamond-shaped; the example presented by Lieutenant Slater in 1893 had a black-and-white checked border. Farther west in Micronesia he also collected a fan-shaped hand-screen in the Gilbert Islands (now Kiribati). On Nukufetau in the neighbouring Ellice Islands (now Tuvalu) Mr. C.F. Wood collected a spatula-shaped fan; this had a wooden handle and probably belonged to someone of rank.

Another fan with a carved handle is in the Horniman Museum and comes from Mangaia in the Cook Islands. It is kite-shaped and has a small 'W' notch on the tip; the wooden handle is carved with the *tiki-tiki* pattern. A humbler version with a plain handle, but much larger (32½in/77.5cm long) and with a plaited carrying-thong, was sold at Christie's in 1979. The Horniman Museum has a Samoan fan with triangular leaf and zig-zagged upper edge, and also a Tahitian fan, long and triangular with a kite-shaped wooden handle with a hole in it.

These palm leaf fans are extremely rare because they rot. Although many ethnographic collections have the carved handles from the royal fans, when the fans survive with their handles we have some of the most valuable fans in existence. In the Polynesian Marquesas Islands, the fans are triangular and three examples with carved handles have been sold at Christie's; the last, the cheapest, having an ivory handle made from the remains of a harpoon head.

The most valuable South Sea fan was the Rev. John William's, sold at Christie's in 1979 for £38,500 ($63,525). Not only did it have most of its leaf and the rarest of handles—the Janus figure—but it had a celebrated history. Williams of the London Missionary Society was

Above: In this varied group there are two Oriental fans: centre top is a Cantonese brisé fan, c.1820, painted with figures in a boat against a red lacquer ground, and with a frieze of animals against a black lacquer ground, the verso similarly decorated with figures in a garden. 8ins (20cms), sold by Christie's South Kensington in September 1988 for £462 ($762). Shown closed is the Japanese fan, c.1880, with scenes of autumn and winter, which can be seen open in the top picture on page 69: the very beautiful shubayama-work on the guardstick makes this a highly attractive fan, even when closed. Other fans in this fine assemblage include 'The Lady's Looking Glass', a printed fan forming a puzzle by having a letter on the front partly written in symbols and hieroglyphics, with the solution on the back (English, late 18th-century); and a scallop-leafed fan painted in the manner of Pillement (the presence of Chinese children and figures shows the importance of the Oriental influence on design) (European c.1750).

told of the existence of the island of Rarotonga when preaching in other Cook Islands, but was unable to find it. At the same time the islanders had heard of Christianity from a Tahitian convert. One of their kings, Makea, set out to find the missionary. He found him in 1823, a disappointed explorer, and was able to set him on the right course and later help him in converting the islanders. In order to show Christianity's superiority, the various kings burned their idols and feasted off the food cooked in the ashes, and so only the fans and handles survive. It is thought that Makea gave Williams this fan.

In this case the price was due to the rarity of anything surviving from this particular island; another fan or fly-whisk handle from Rarotonga also fetched £38,500 ($63,525) in 1979, despite having neither its fan nor the fascinating provenance of the other piece.

From the Banks and Torres Islands of Melanesia come

oblong flag-shaped fans of split and plaited palm leaf.

In other South Sea islands fans are of lesser importance. However, in Fiji, a ping-pong bat-shaped fan of decorated palm leaf with string handle was collected by Lieutenant B. E. T. Summerville in 1896 and presented to the Pitt-Rivers Museum, and the British Museum has a pierced wooden fan from Tonga, illustrated in *Fans from the East* (plate 61).

In India, fans were less important but (not surprisingly in a sub-continent) they show great diversity. Fine hand-screens of plaited ivory from the late eighteenth-century were exhibited in *Fans from the East*, and another example was sold at c.s.k. A large number of Indian fans are side-mounted and revolving.

In the Naga Hills near the Burmese border fans are plain; in Indore they are of woven plaited straw. In Peshawar (now in Pakistan) they are axehead-shaped and embroidered. Some fans from Bombay are decorated with beads and tinsel, others are of cotton embroidered with sequins and beetle wings. From Madras, on the opposite coast, come some very curious fans. One, over three feet high, in the Pitt-Rivers Museum is made from bussa palm; the handle is formed by the thicker part of the stem, and where the stem becomes the leaf, it is twisted to the side and plaited.

A fan from Madras in the Horniman Museum is large, almost circular, and made of bamboo painted with peafowl: the handle is of lacquered wood and has a ring-

Left: Indonesian, c.1850. The folding leaf is made of shaped buffalo hide panels painted with dancers. The sticks are of wood.

shaped insertion at the bottom of the fan. The fan was originally edged with peacock feathers. A much simpler example is in the Pitt-Rivers Museum. A fan from Calcutta in the Horniman Museum is circular, made of paper decorated with straw-work and fringed with dried grass.

In neighbouring Burma large fans are made of pleated bamboo. They are leaf-shaped and have curved handles mounted diagonally—the one in the Pitt-Rivers Museum was collected by the Reverend A.H. Finn in 1888. A simpler example of palm leaf with a plain wooden handle is in the Horniman Museum; these fans screened the priest's face when he preached to women. Other fans were made in Burma for export, including folding satin fans—one painted with a procession was sold at C.S.K. recently—and also brisé fans folding to form a parakeet, as illustrated in *Fans from the East*.

To the west in Persia, leaf-shaped hand-screens were made of papier-mâché and painted like the pen boxes one often finds there. Similar fans were made in India under the Mughals and later under Persian influence.

In Africa (as in South America) there is great diversity, although fans are less common than fly-whisks. These latter predominate in South and East Africa and exist in West Africa as they do in India and elsewhere. From Madagascar, off the east coast, come grass-work fans, and farther north, in the Arab settlement of Zanzibar, revolving fans were made with axehead-shaped mounts profusely covered with brightly coloured beadwork.

Above: Cantonese export style (but perhaps European), c.1830. A cabriolet fan, the upper leaf painted with figures in both indoor and outdoor scenes: the figures have ivory faces. The lower leaf is painted with birds and flowers. The reverse has more figures on the upper leaf. The sticks are of tortoiseshell, carved and pierced. 10³⁄₄ins (27cms).

From West Africa come circular or leaf-shaped fans on wooden sticks. The Pitt-Rivers Collection is rich in Nigerian fans and includes two hide fans, one from Kano in the north and the other from Lagos in the south—used by the wives of chiefs to fan flies from their husbands; also a circular raffia fan from Ikot Ekpene collected in 1932 by J.F. Ross. A semi-circular fan in the Pitt-Rivers Museum, given to them in 1913, is applied with cloth decorated

Above: Huron, late-19th century. A feather handscreen, the handle worked in moose hair on birch bark.

Below: Two 19th-century Indian handscreens, decorated with beetle wings.

with a trellis of braid, held in place by studs edged with ostrich feathers. Nigeria also produced interesting coloured woollen flag-shaped fans. The Horniman Museum has two hide cockade fans, as well as a leather fixed fan of fan shape trimmed with feathers to make it over three feet wide, and from the Cameroons, Nigeria's immediate neighbour to the east, a beadwork circular fan with a large feather fringe.

The Pitt-Rivers Museum also has two other West African fans of typical type. One, a great curiosity, is of wood carved all over and probably used in ritual dance; it is probably well over 100 years old as it was transferred to the Pitt-River Museum from the Ashmolean in 1886, having arrived earlier from the Christie Collection. The other is probably not so old, but both leaf and handle are of black and white cow-hide and the leaf-shaped mount is applied with motifs of green and white leather.

The North American Indians mainly use feather fans, sometimes swan's wings, but a bat-shaped fan in the Horniman Museum was decorated with porcupine quill-work flowers and originally edged with rushes.

In the hottest parts of South America birds are abundant and there fans are made of feathers, but in the Cayapas River district of Ecuador, lozenge-shaped twill-work fans were used to feed charcoal fires, in the same way as Spanish and Portuguese twill-work fans were used.

The only European ethnic face fans are the Madeira cockade fans of fine dried grass.

Of course, this brief survey only outlines the types of fans found in Africa, India and the South Seas: the enthusiast should direct his or her attention to the many books on individual tribal cultures.

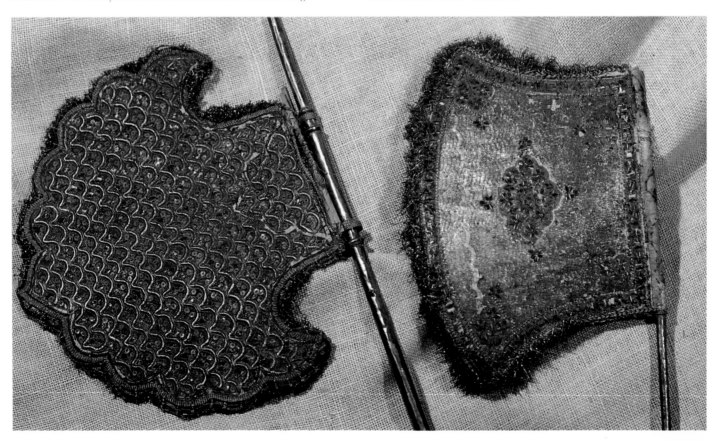

NOVELTY AND FEATHER FANS

On the whole, novelty and trick fans appear in greater quantity during the later part of our period. These fans could be works of art but were intended as amusing presents rather than lasting treasures.

T rick fans are a popular collector's speciality—from the simple but enchanting double-image fans that show two different pictures depending on the direction in which you open them, to articulated fans that slide contrasting images up the leaves through levers on the guardsticks; from fans with spy holes, to 'telescopic' fans that expand to nearly double their length, or fans that show different pictures when held up to the light. There are many fascinating novelty fans of unusual size and appearance—cabriolet, giant, miniature, parasol, cockade and feathered.

DOUBLE-IMAGE FANS

The earliest novelty fans are the double-image fans. There is a fine example in the Musées Royaux d'Art et d'Histoire, Brussels (1768–1944), painted with classical scenes and flowers on a dark ground with tortoiseshell sticks, probably French, *c*.1700. A much simpler and later example painted with figures in a landscape (*c*.1700) was sold at C.S.K. from the Baldwin Collection. As one opens the fan the normal way, one sees a lady and gentleman sitting in a garden; opening the other way reveals a moneylender, and on the reverse a vignette. Some fans conceal four separate views, such as the parasol fan in Wisbech Museum, or a similar one in Munich; these are described more fully under Parasol Fans: the trick is that each stick supports a double fold and a different section is extended depending upon the direction in which the fan is opened.

This method has also been used to conceal risqué scenes, known as a *double-entente* fan: there is one of this type in *Fans from the East* (plate 1). When opened in the normal manner to the right, this late nineteenth-century Chinese

Above: French, c.1770. A double-image fan, one leaf painted with a proposal scene of an elegant lady holding a bird cage and her lover, with a canary on a blue string, at the side of the Altar of Love: the hidden scene depicts a lady falling into a trap watched by agitated figures. The reverse is painted with sprays of flowers, the ivory sticks decorated with red and gold foil, and silvered. 11ins (28cms).

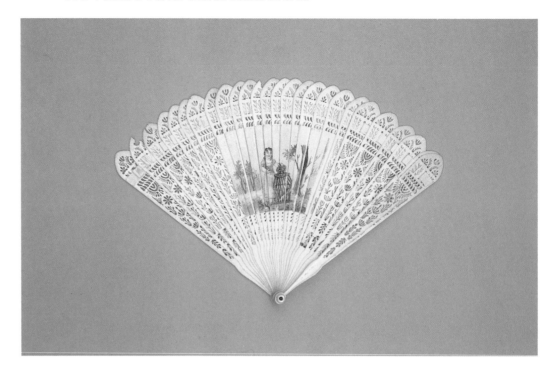

Right: c.1815. A double-image brisé *fan, the ivory sticks pierced and painted with four different vignettes: a woman with a dog, feeding chickens; a little girl with a large cage or lobster pot (this image seen here); an urn of flowers, and a basket of flowers.*

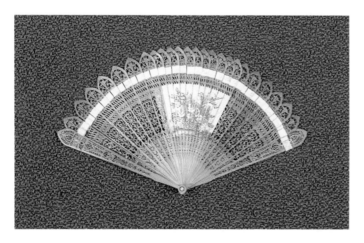

Above: c.1815. A pierced horn double-image brisé *fan, painted with a basket of flowers and other vignettes. 6¹/₂ins (17cms)*

fan depicts a lady and attendants in a garden, but when opened to the left, it reveals titillating scenes (as does c.s.k., 31 January 1980, lot 85).

Brisé fans, too, can be double-image fans. In these the blades are deceptively wide and overlap halfway so that half the picture is painted on each half of the stick, and again, depending upon which way the fan is opened, they show a different scene. They first appear in England on sandalwood brisé fans of the 1780s and 1790s. The scenes are often applied stipple engravings and they are usually threaded with pretty green and pink striped ribbon. Then, in about the 1820s, a charming group of pierced ivory or horn brisé fans appear, probably also English; they have pretty, brightly painted fancy scenes on a plaque in the centre of the fan, such as baskets of flowers, a huntsman, or children. Like the foregoing fans, they closely resemble other normal fans of the period and so are very much trick fans which can easily deceive the inexperienced collector, who might pass them by undiscovered.

ARTICULATED

The next group of trick fans involves the sticks rather than the leaves. In about the 1770s there is a very rare group, probably produced in Germany or possibly France, with articulated sections in the guardsticks: a metal lever is cunningly concealed in the *rocaille*, and when this is pushed up or down the sections move up and down to show different faces or, if articulated, figures wave or offer bunches of flowers. One such fan in the Victoria and Albert Museum has three female heads which are exchanged for three male on a lacquered slide, possibly the same fan as that exhibited in the Karlsruhe Exhibition in 1880. (Collection of H.R.H. Erbgrossherzogin Pauline von Sachsen-Weimar). Another example was sold at c.s.k. from the Baldwin Collection: an articulated fan celebrating the birth of the Dauphin (1781); the levers raise a figure of Cupid and the crowned escutcheon of France.

There is also a trick or puzzle fan, possibly Chinese, c.1780, where the ivory guardsticks have little channels or tracks carved in them, slotted with ivory balls which run up and down as the fan is tipped.

CARVED STICKS

Another group is of the Rococo period and is mainly Spanish and Portuguese; these have handles carved to form elaborate shells, sunflowers and other novelties when closed. This, in the case of Spanish fans (or rather fans for the Spanish market) and Portuguese fans, involves the carving continuing down the side of the closed fan like a fore-edge painting.

SPY HOLES

In some versions spy holes are the eyes of a mask. There is a group of interesting printed fans, c.1740, of which several are known, including one in the Baldwin Collection; another is in the Boston Art Gallery and the third in a private collection. These show a face surrounded by

Above: French, c.1781. An articulated fan celebrating the birth of the Dauphin, the silk leaf painted with the Dauphin in a cradle guarded by an angel, flanked by vignettes of putti riding dolphins and busts of his parents. The reserves are painted with flowers and dolphins, decorated with ribbon-work and spangles, and inscribed 'Vive le Roy la Reine et Monseigneur le Dauphin'. The ivory sticks are carved, pierced, and painted with the Dauphin and his nurse, fleur-de-lys, and dolphins. The guardsticks are set with mirrors and carved with dolphins, concealing levers which control the moving parts: when manipulated they raise a figure of Cupid and the crowned escutcheon of France. 11ins (28cms).

views from Spanish life, including a fan shop and a music shop. As two of these fans come from America, one could suggest that even though there was an English export trade in fans to Spain, they were made for the Asiento, or annual ship to Spanish America, established by the Treaty of Utrecht.

From about 1760 there appeared fans with net spy holes set in the leaf, partly decorative and partly a useful novelty. Spy holes are usually disguised as part of the design, as in some of the Swiss fans made at Winterthur. A Victorian version was sold at Christie's in 1974. This was a masquerade fan, the net leaf with a masque, the wooden guardsticks set with a mirror and an *étui* containing scissors, a boot-hook and a pencil (by W. Thornhill & Co., 14½in/36cm, *c*.1880).

There is another equally well-fitted fan but without the masque, patented by W. Thornhill, *c*.1895, and illustrated (plate 42) in Mrs. de Vere Green's book. The thick wooden guardsticks contain secret compartments, with a comb and manicure implements in one side and sewing tools including scissors in the other, and the thimble in the tassel hanging from the rivet. So handy for running repairs! The photograph also shows an ivory brisé fan with mirror and powder compact in one guardstick. One often finds a looking-glass on guardsticks of fans—particularly Spanish lithographic fans of the mid-nineteenth century. Rimmel's Cassolette Fan has a container for rouge on the guards (cat. No. 217, *The World of the Fan*, Harris Museum, 1976). A rare eighteenth-century fan sold at Sotheby's Belgravia in 1979 had a secret panel in the guardstick.

To revert to spy holes: there is an example of spy holes in the sticks of an 1830 stipple-engraved fan with mother-of-pearl sticks (see Mrs. de Vere Green's book, plate 39). Later on, in about the 1820s, spyglasses were sometimes set in the pivots of fans, often brisé, sometimes parasol brisé fans. Later in the nineteenth century there is even an instance of a pair of opera glasses attached to the rivet of a fan (see Mrs. de Vere Green, plate 42). No. 225 in *The World of the Fan* exhibition at the Harris Museum is a similar fan (*c*.1885).

In the 1760s to 1780s a spate of fans appear with novelties in their pivots or on their guardsticks, such as the superb fan of *c*.1755 which has a jewelled watch by Upjohn of London, signed No. 248, as a pivot (sold by Christie's and exhibited at Bruchsal in 1989), and a similar fan from the collection of the Dowager Marchioness of Bristol illustrated by Woolliscroft Rhead (plate 153). The guardsticks on these fans are also jewelled, as they would have been used at court—the very grandest court fans of that period had jewelled guardsticks since, in the presence of royalty, it was customary to hold fans closed, and thus the jewelled guards presented a very grand appearance. No. 227 in the exhibition *The World of the Fan* was an Edwardian version of a watch on a fan: an ostrich-feather fan with a watch on the guardstick.

There is an amusing fan in the museum at Bordeaux with a thermometer set in the guardstick to test one's ardour (like those modern novelties, 'passion testers', for which one clasps a phial of coloured water which changes colour if the hand is hot enough); a similar fan was sold at c.s.k. in 1979.

TELESCOPIC

Telescopic fans first appear in the middle of the eighteenth century. These are fans with paper leaves only loosely tipped on to the sticks. They have to have thick double mounts, so their slightly bulbous appearance may give them away. Examples include one sold at Christie's in 1973, a telescopic fan painted with a fairground scene and with ivory sticks, 6½in (16.5cm), expanding to 11in (28cm) (French, *c.*1760); and another (lot 93 in the same sale), in its original short box by Clarke, who on later labels claimed to be the inventor of these pocket sliding fans.

These telescopic fans are normally rather crudely painted, with plain ivory sticks. To fold them when lightly closed, one holds the handle and gently pushes the leaf down (very gently, as over the years they have often come out of true); the leaf then comes down to cover about half the length of the sticks and produces a neat closed fan, rather bulky, but of about the length of a small brisé fan that can easily fit into a reticule for travelling. These have sometimes appeared with their original short boxes, which is a quick way of identifying them. One such fan must have fallen into the hands of someone who had not understood it and had glued it up to its extended state.

Above: French, c.1900. The leaf is trimmed with silk flowers, and the wooden sticks fold in a circular fashion so that when it is closed the fan forms a bunch of violets. Signed with a partially legible address, possibly Av. de l'Opera, it is believed to have been purchased at the Paris Exhibition in 1900.

Below: English, 1778. The leaf is painted with personifications of the Arts on the front, and on the reverse with a tree in blossom. The ivory sticks are pressed with further representations of the Arts. The guardsticks are elaborately carved, and inscribed 'Duflos fesit, London 1778': the carving is in high relief, and the devices are putti, globes, and other astronomical instruments. One of the guardsticks is set with a Celsius thermometer. 11ins (28cms).

NOVELTY AND FEATHER FANS

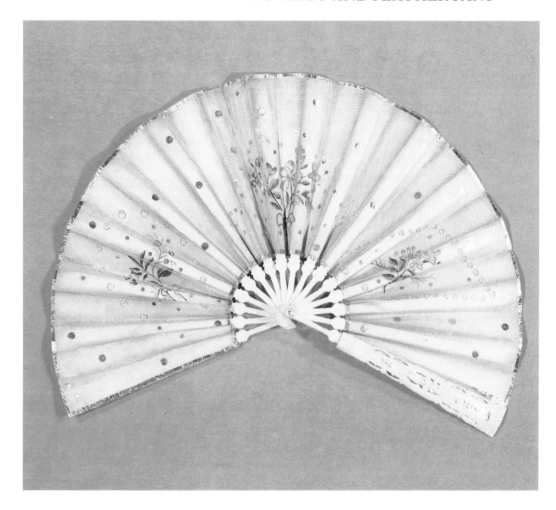

Left: This illustration is of a telescopic fan, open, but not extended. In this state the leaf covers about half the length of the ivory sticks, and when closed forms a very compact item for travelling. Telescopic fans are generally more crudely painted than other types, and normally have plain sticks. This textured example is decorated with a simple painting of flowers.

Below: c.1785. The chickenskin leaf is applied with a hand-coloured stipple engraving showing a charming scene of two little girls, one sleeping contentedly while the other steals her hat, and a kitten playfully pawing a bird cage. The ivory sticks are pierced.

Lithographic versions of telescopic fans dating from the 1840s also exist, as do plain linen ones of a little later, and there are examples which have strayed from their turn-of-the-century fitted handbags.

CABRIOLET FANS

In the 1760s cabriolet fans were fashionable in France. These were named after the little English cabriolet carriage in vogue at the time, as their double mounts and sticks resembled the spokes of a section of a large cabriolet wheel with its reinforcing circle across the spokes. Some very fine examples exist. A cabriolet is often incorporated in the design. One fan sold at C.S.K. in 1976 also had a fore-edge painting of bagpipes, extremely rare in fans. Cabriolet fans are much sought after by collectors. They disappeared from fashion with the cabriolet, but reappeared in the nineteenth century, when there are examples from Canton and some lithographic fans.

In the late eighteenth century, there are some rare French Royalist fans. One is painted to show Cupid with a magic lantern throwing on a screen a picture of a pansy, which, when held against the light, shows concealed portraits of Louis XVI and Marie Antoinette and the Dauphin, the pansy being a transparency (see Karlsruhe Exhibition catalogue, ill. 13: collection of Frau E. Fuset). There is a similar fan in the Schreiber Collection, 'Le Songe': a woman sleeps by a tomb in a rocky cemetery and again, if one holds the fan up to the light, a figure of Louis XVI

stands on the grave. One of the Bastille fans in the Schreiber Collection has wooden guardsticks carved and pierced as a simplified version of the Bastille.

In the 1780s, there were some more brisé fans carved in the form of a shotgun. C.S.K. sold an example in ivory. In about 1820 an ivory brisé fan was carved to close in the form of an arrow.

Two amusing late nineteenth-century fans were sold at C.S.K. in 1977 and Sotheby's Belgravia in 1979. The former has sticks that fold up to become a policeman, and the latter a clown. The policeman opened to show a policeman's shadow on an orchard, and numerous boys fleeing over a wall; the reverse showed the other side of the wall and the boys tumbling off. A similar idea occurs on certain Chinese eighteenth-century spy-hole fans where the spy holes are windows and their pictures show inside and outside views. This amusing idea also occurs on other late nineteenth- and early twentieth-century fans; there are some good examples by Duvelleroy in the Victoria and Albert Museum.

Erté (Romain de Tirtoff, 1892–1990), designed some strange fans, including two which appeared in *Harper's Bazaar* in February 1922; they are both disguised as long tassels to hang from the wrist—one opens to become a brisé fan of lacquered wood with gaily coloured silk strands hanging from it, the other is of jet beads. The tailpiece to the foreword of his autobiography *Erté Fashion* (1972) is a feather parasol fan that doubles as a parasol when the feathers are swivelled to a horizontal position by pressing a lever in the long handle.

From the 1780s until about 1800, printed fans were the main source of many novelty fans, for example 'L'Oracle', the horoscope fan with an ivory arrow-shaped pointer, and several with conundrums, one being published by Sarah Ashton in 1794. Such fans were also produced in France and with rebuses; there is one with a hand-coloured etching, 'Charades Nouvelles', engraved by Benizy.

Jokey fans such as 'Fanology' (1797) fascinated late fan collectors; Duvelleroy even published a leaflet called 'The Language of the Fan' (see Gostelow, fig. 43), and Addison in the *Spectator* (102, 27 June 1711) satirically describes an imaginary academy for training the young in the exercise of the fan. Christopher Towle, an Oxford dancing master who published an etiquette book, made great emphasis on keeping the elbows in. Taken within reason, these suggestions make sense, as it is important for fans to be handled gracefully. The present generation is, of course, out of practice!

Some fans also have amusing details on their sticks, such as two charming Flemish fans, their ivory sticks painted with amorous mottos and vignettes; one was sold by C.S.K. from the Baldwin Collection; the other is in the Musée des

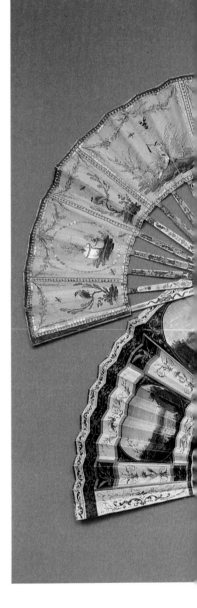

Right: Among this stunning array are two rare articulated fans. Top, far right, is a French fan, c.1775, which was sold by Christie's South Kensington in May 1989 for £5500 ($9075). The silk leaf is painted with an allegory of a milliner's shop, incscribed 'La Folie l'a invente et la Mode l'a adopte', with a jester giving a hat to the Milliner while a cleric arranges another hat on an elegant customer watched by Cupid. The reserves are embroidered with spangles and ribbons, and the ivory sticks are pierced and silvered. The guardsticks conceal the articulated devices. In the one that can be seen a young man kneels at the Altar of Love: as the lever is slid down Cupid arrives with a garland of flowers. On the other guardstick the movement pushes a parrot out from behind an urn to peck at a lady. The other articulated fan is shown next to the first, closed. The mechanism can clearly be seen: when manipulated the sportsman moves. On the other guardstick a lady lifts a mask to her face. This dates from c.1770.

Arts Décoratifs and was exhibited at Bruchsal in 1989–90.

PARASOL AND COCKADE FANS

Wisbech Museum possesses a very rare double-image parasol fan; it is probably Flemish and dates from the early eighteenth century. The divided linen leaves are painted in sets as follows: the first with crippled beggars in bright colours; the second with scenes of gods and goddesses, also in bright colours; the third with Cupid and other figures standing in individual landscapes and hillocks and bearing symbols of gods painted in monochrome in brown; the last with similar figures in monochrome in blue bearing symbols of the Arts and the vintage. The sticks are ivory, the thick guardsticks of pierced and painted wood, one set with a circular mirror. The Victoria and Albert Museum (to whom it is, at present, on loan) suggest that it may have been a Christmas or birthday present to someone born at that time of year, as poverty and disfigurement are symbolized by Saturn, the ruling planet. There is a very similar fan in Munich.

Above: Mid 19th century. A parasol fan, with silk leaf and ivory handle.

In about the 1880s there is a group of very cheap-quality cockade or parasol fans, sometimes only opening to a semi-circle, and also known as dagger fans, as the cockade is contained in a spatula-shaped wooden sheath. There are cords at each end. To open the fan, one pulls the top cord and it will snap up and unfurl; to close it again one pulls the bottom cord and the fan slips back into place. The case is often inscribed 'Ricordo' and set with a looking-glass on the side. In January 1980 C.S.K. sold a much finer example (c.1840).

Another type unfolds from velvet- or American cloth-covered sticks, rather like certain Christmas decorations or an old-fashioned pair of nutcrackers in reverse.

Multiple parasol fans also occur. One with as many as four parasols or cockades on sticks of decreasing size was sold at C.S.K. in 1979; this is said to have come from Zululand.

There are other imaginative types of parasol fans. One, by B. Days, Snow Hill, Birmingham, folds into Gothic candlestick-like ornaments decried by A.W.N. Pugin in *Contrasts*. There are also some late eighteenth-century ver-sions with green silk, curved wood and ivory handles, and Canton ones of pierced ivory that sometimes come with their own hanging boxes (see Mrs. de Vere Green's book, plates 15 and 16).

C.S.K. have sold several French fans decorated with cloth flowers, which fold in a circular manner to form posies.

FANS OF UNUSUAL SIZE

From about 1870 until the late 1880s Paris was producing expensive bisque-headed dolls complete with trousseaux of adult clothes in trunks. These trousseaux usually included a tiny fan, either ivory or bone brisé or satin, with painted ivory or bone sticks about an inch or so long. There also exist some slightly larger fans, often litho-graphic, dating from the turn of the century, which, being miniature, are said to have been made for children. They are often by Duvelleroy and other well-known makers. Perhaps they were given to clients' children, encouraging their mothers to buy the more expensive fans. These minia-ture fans are all fairly rare and much sought after by collectors. For the former type, fan collectors, of course,

Above: 1879. Eventail à dislocation, a trick fan, the leaf composed of pink and olive green ribbons which when opened the wrong way (from left to right) fall apart. The fan has wooden sticks. 14ins (36cms).

have to compete with keen doll collectors. Small fans for children are also thought to have been made in the eighteenth century.

Giant fans, about 16in (40.5cm), were produced in France in the 1740s and 1780s, mainly of decorated paper, though a few were printed; they all had wooden sticks, occasionally painted. They appeared again about the 1880s, and were sometimes used as fire-screens by the Aesthetes (see ill. of Norman Shaw's house in Mark Girouard's book of *Sweetness and Light*). There is another in the collection of

The Queen at Osbourne House: 27in (68.5cm) long (58in/122cm) diameter, it was painted with roses by The Queen's granddaughter, Ella of Hesse (see Gostelow, fig. 80). There were also giant Japanese fans produced at the time, including some sold at C.S.K. There is a giant fan in the Messel Collection, and two others appear in a late nineteenth-century photograph of M. Hervé Hoguet's eventailliste's shop (Mary Gostelow, fig. 97).

Another small type of fan is the dance programme fan, made for specific balls such as Mr. Foster's dance at Hornby Castle in the 1880s. These were sometimes ivory brisé fans with a pencil slotted into one guardstick. There is also an example in Letchworth Museum dating from *c*.1860, 2½in (6.5cm) long, with a corkscrew of metal on one guard holding the pencil; a gilt chain and ring are attached to the pivots to hang from the owner's finger whilst dancing. When admirers asked the owner to dance, either they or she signed the blades in advance.

At about the same time, brisé fans were often decorated with autographs, sometimes even with inscriptions, photographs and illustrations as well, and other brisé fans were decorated with collections of crests cut from letter headings and envelopes applied to the sticks. Another novelty from the 1870s was the Calendar Fan, a printed paper brisé fan. Dr. van Eeghen has a chocolate box in the form of a fan, from P. Nieuwerkerk & Fils, The Hague.

A most unusual fan was included in the Victoria and Albert Museum's 1974 travelling exhibition of musical instruments. A parasol fan, 12in (30.5cm) long – the handle formed as a violin that plays – it belonged to M. Louis Clapisson (1808–66), Premier Conservateur of the Musée Instrumental of the Conservatoire.

Other novelties include Japanese daggers disguised as fans, brisé fans made of Welsh slate (Welsh Folk Museum), and an American tinware hand-screen fan moulded with

Below: American, late 19th century. A Scrap fan, the satin leaf decorated with American crests and letterheadings, including Bryn Mawr, Boston Art Club, Yale, Pullman Palace Vestibuled Train, and the Knickerbocker Athletic Club. It fetched only £10 ($16.50), because torn, at Christie's South Kensington in February 1990.

Above: This butterfly fan is 23ins (58cms) wide and 18ins (46cms) long. Its wings are made of gauze, painted yellow and orange, with mauve veins and some gold and silver touches. The handle is of black tin.

pleated leaf, probably from New England. Another plain folding fan opens normally from left to right, but when handed to an unsuspecting person who opens it the other way, it falls to pieces. The secret is in the folding of the mount—this is made of as many pieces as the sticks and these are folded to engage with their neighbour when opened to the right, but when opened to the left the folds do not engage and fall apart. An American manufacturer advertised such fans for 20 cents in the 1880s. A modern version from Bali is illustrated in *Fans from the East* (see plate 45).

The Craftsman of 1784 advertised Edward Vaughan's invention of the Necromantic Fan or Magic Glass Fan. Mrs. de Vere Green also discovered another rare type. An example of this is illustrated in *Fan Leaves* (The Fan Guild), plate XXIV. This is a Panorama Fan (French, *c*.1830); a rectangular hand-screen with a central rectangular gap holds the panorama. This is rolled round ivory-handled spindles arranged vertically within the thickness of the fan. Small turned knobs protrude below the lower edge of the fan at either side of the handle; they draw the pictures to and fro, depending on which way they are turned.

Yet another unusual type is the revolving or roll-up fan of Japanese origin, known as *Maki uchiwa*. A slender bamboo handle is slotted for most of its length and into this slot fits the circular mount, consisting of a large number of slender bamboo strips glued to a paper or silk backing in such a way that the mount can be rolled up like a blind. When the mount is slotted in the handle it is secured by

a central pivot. By rotating the fan 90° it can be rolled round the handle and tied down.

Another unusual fan mentioned by Mrs. de Vere Green is made of hard black rubber. When the handles were rotated it opened to a circle, 9½in (24cm) diameter. This was patented by Henry B. Goodyear, the tyre magnate, in 1858. The fan had nine leaves. An example survives in the Doble Collection, Boston. Mrs. de Vere Green also illustrated (fig. 48) 'The Pedal Zephyrion' (*c*.1878), a fan attached to a rod, on pedestal-like and standard-lamp-like protruding pedals. The lady reader sitting beside it is able to fan herself by pedalling.

A delightful English invention has survived: 'The Butterfly Fan', provisional patent No. 14726. Three examples were sold at Christie's South Kensington in 1979. The gauze wings are mounted on wire frames, the handle forms the body, with a lever which closes the wings when pressed.

FEATHER FANS

In about 1880 feather fans returned to fashion. Painted goose-feather Canton fans had, of course, been on the market in the 1830s and 1840s but feather fans had not been really popular in Europe since the seventeenth century. They remained in use until the 1930s as Court presentation fans of curled white ostrich feathers. In the 1890s feather fans often consist of simulated birds' wings.

In the twentieth century more and more ostrich feathers were used. They were imported from South Africa, curled and mounted on tortoiseshell, mother-of-pearl, simulated amber, ivory or wood sticks, depending on the quality of the fan. Many have survived in pristine condition in huge satin-lined fitted boxes by Duvelleroy and others. In the 1920s and 1930s they were dyed beautiful colours, sometimes shaded or streaked. Sometimes extra feathers were attached nimbly by hand to the tips of individual feathers, making the fans immensely long and producing a 'waterfall' effect. Sometimes on very expensive fans the mother-of-pearl sticks have been tinted blue, pink or yellow to match the dyed feathers. The owner's initials have often been attached to one guardstick in silver, gold and occasionally large diamonds.

There were also some small tortoiseshell brisé fans tipped with peacock and other feathers from about 1890. At the lower end of the market, Sears, Roebuck & Co., a Chicago mail-order store, advertises in its 1902 catalogue:

> No. 18R 989. Genuine ostrich feather fan, new size, made of fine quality stock, white enamelled sticks. You will find this a rare bargain at the special price we quote. Colours cream or black—Price, each 83c. If by mail, postage extra 5 cents.

This fan appears to be of a quality inferior to many of the fine examples by Duvelleroy and others that have appeared on the market recently. In England, the Army and Navy Co-operative Society advertised 'real ostrich feather fans' in their 1924 catalogue, as mounted on a single handle, 9/- upwards, with 5 sticks for 24/-, and 10 sticks for 75/-.

On some of the cheaper feather fans, the feathers have been applied as individual rosettes or cockades to each stick.

Another type of feather fan that appears on the market is a superbly made hand-screen of gaily coloured feathers surmounted by a stuffed humming bird. These were made in Brazil towards the end of the nineteenth century, and are often in their original labelled cardboard boxes.

Above: *Brazilian, late 19th century. A handscreen decorated with feathers and stuffed hummingbirds. The handle is of ivory.*

Right: *A gorgeous shaped handscreen, probably made in South America for the European market, 17th or 18th-century. The handle is of white metal, and the screen itself is of catgut, stained blue, appliqué with featherwork birds holding sprays of flowers, the reverse similarly decorated. The feathers used to decorate the fan are from birds found in the northern half of South America, including green feathers from the Quertzel and purple feathers from the Cotingas. 13ins (33cms), sold at Christie's South Kensington in April 1986 for £2700 ($4455).*

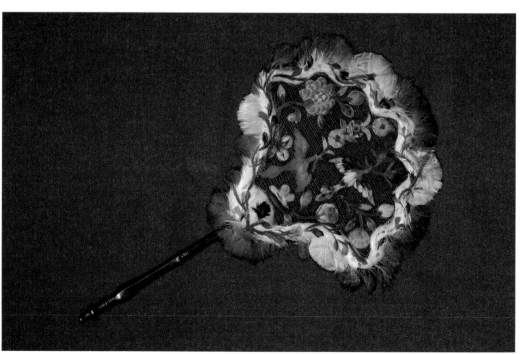

CARE AND MAINTENANCE

Once you have started to collect fans, there are

certain questions you will ask about looking

after them and when to restore them.

Twenty years ago, even ten or five years ago, collectors were careful to only buy fans that were in good condition. Serious collectors have now realized that the early fans are getting rarer and rarer to come by, and I now find that slight damage does not affect the value of fine fans nearly as much as it used to.

As with other works of art, everyone has a different idea of how much or how little conservation should be carried out on their object, so it is best to leave conservation work to the next owner. However, there are some remedies you could carry out yourself—for instance, two fans in the 15 May 1990 sale at Christie's South Kensington arrived without their pivots or rivets—presumably these once incorporated rose diamonds, and some unscrupulous person removed them for re-use. For £4 (about $6) each I was able to replace these on behalf of the vendors from the supplies of the Fan Museum in London. This was well worth while—one of the fans, a Duvelleroy fan painted with a cat, went on to fetch £900 (about $1,500).

Like watercolours, fans should be kept out of the light and at an even temperature—it is unwise to display them open for longer than a month or so at a time, as they begin to warp. Fans should be stored in an acid-free environment, possibly in boxes. If you plan to display your collection in fan-shaped frames, they should be backed with acid-free card, and the fans should be removed every two to three months. It is preferable to sell fans unframed—in fact, there is usually a 20% discount on fine fans sold framed, although cheap decorative lithographic fans in attractive fan-shaped cases do tend to sell well.

Clean breaks in sticks might be fairly easy to repair. Collectors could consult a museum on which glues to use, but it would be wiser to send them to a trained conservator for repair, to avoid any strain on the fan leaf. Many damages can be repaired by a skilled craftsman: torn paper can often be backed with fine tissue, and stains from Sellotape can be bleached out, though this often causes the fan to fade. Gold and silver on sticks can be polished. Oxidation of body colour can occur when the white in the paint goes black. The ribbon threaded through brisé fans is often missing—The Fan Circle International has supplies of the fine ribbon required, and it is possible to learn how to repair these. However, if traces of the old ribbon or paper remain, these should be retained rather than replaced, as the original ribbon is more valuable and usually more attractive. But in general repair work is all very skilled, and should not be attempted by an untrained hand.

One almost impossible damage to tackle is 'shattered' silk—as found on *fin de siècle* fans, where the salt lead applied to make the silk more lustrous has destroyed the fibre; new silk gauze would have to be laid down, which is not normally as attractive.

The fan museum in Greenwich, London, has just set up a conservation department; different specialists cover every type of work needed, as fans are made up of so many different materials—paper, silk, lace, leather, feathers, ivory, tortoiseshell, mother-of-pearl, wood, silver and gold, to name but a few. Remember that the trained conservator has learned the chemistry of the pigments and the materials, and has examined numerous fans. So when in doubt, call a professional.

GLOSSARY OF TERMS

à disposition: woven or made to shape

à grand vol: a fan that opens to 180°

assignat: a bond issued by the French Revolutionary administration against the value of Church lands

battoire: a folding fan with particularly wide sticks, resembling a tennis racket

body colour: watercolour mixed with chalk

brisé: a fan comprising of sticks only

cabriolet fan: a fan with two leaves, one above the other, similar to the section of a wheel of a cabriolet, a fashionable carriage in the 1780s

canepin: fine parchment, *see* chickenskin

catgut: a strong net woven from animal guts – sometimes used for spy-holes of 18th-century fans

celluloid: an early thermo-plastic

chickenskin: very fine kid, or parchment, said to have been made from aborted foetuses

chinoiserie: a European imitation of the Chinese style

church fan: a fan painted with a subject suitable for display in a church, mainly used in Holland

cloisonné: a form of enamel-work

clouté: inset with semi-precious stones or mother-of-pearl

cockade: a fan that opens to 360°, usually with guards that extend to form handles

Comedia dell'Arte: the traditional provincial Italian pantomime with Punchinello, Colombine and Harlequin

découpé: paper or vellum with holes punched in it to resemble lace

domino: 18th-century masque fan and cloak worn at masquerades; the two known examples were sold by Christie's

double-entente fan: double-image fan with a smutty subject on the hidden leaf

double-image fan: fan that shows a different scene or scenes when opened left to right

double-sided fan: a late 19th-century fan, usually with gauze insertions, the verso showing the scene seen from behind

en camaieu: monochrome, painted in shades of a single colour etching: a method of engraving that produces fine pen-like lines

Éventailliste: French fan maker

flabellum: early Medieval fan used to keep the flies off the Host during Mass

fontange: a fan taller in the centre than at the guard-sticks, resembling the lace headdresses of the 1680s

gorge: part of the fan immediately above the pivot

guardstick: the outer sticks of a fan, often more decorated than the rest

hand-screen: a rigid fan, square or circular, mounted on a stick

jasperware: unglazed pottery, usually white figures against a coloured ground; invented by Wedgwood, inspired by the antique; occasionally used on fans

kid: soft leather

leaf: the fabric part of a folding fan – the mount

lithography: a 19th-century method of producing prints

Mandarin fan: Cantonese export fans decorated with figures with ivory faces and silk clothes are often called mandarin fans

mica: a thin transparent mineral found in India

monte à l'anglais: a paper leaf of single thickness, more common in England than elsewhere

mount: *see* leaf

pagoda sticks: sticks carved in clusters in *chinoserie* style

palmette: fan-shaped like a palm leaf, similar to fontange

parasol fan: cockade with a single stick, sometimes lifting to resemble a parasol

pique: dots of gold or silver metal set in tortoiseshell or ivory

pivot: the metal rod that holds the sticks together

recto: the front side of the fan

reserve: the outer portions of the fan leaf

ribs: the upper part of the sticks that support the leaf

rivet: the screw that holds the pivot together, often jewelled

scrap-fan: a fan decorated with scraps, usually crests and letter headings

shubayama: elaborate Japanese inlay including coloured stones and mother-of-pearl, named after the family who were its best exponents

spangles: pressed metal-paper shapes

stipple engraving: a method of engraving using dots

taffeta: silk

telescopic fan: fan in which the sticks slide into the ribs to make the fan smaller to carry

trompe l'oeil: a fan painted to deceive the eye, to look like a real three-dimensional object rather than a watercolour picture

vellum: a stiff parchment much used before the invention of paper

vernis-martin: a varnish invented by the Martin family for furniture and carriages – in fact never used on fans; a similar varnish was used on brisé fans during the first half of the 18th century and in about 1900

verso: the reverse of a fan

vignette: a small painted scene

'waterfall': a large ostrich feather fan where extra down has been added to the tips of the feathers

FAN EXHIBITIONS

1870 Loan Exhibition of Fans, South Kensington Museum, London
1874 Fan exhibition, Milan
1876 Antique fans, Moravian Museum, Brunau, Czechoslovakia
1877 Exhibition of Fans, Liverpool Art Club, Liverpool
1878 Competitive exhibition of fans, Worshipful Company of Fan Makers at Draper's Hall, London
1882 The celebrated collection of fans of Mr Robert Walker, The Fine Art Society, London
1882 The Society of Decorative Art, New York
1885 Les éventails de la collection de M. Emile Duval, Geneva
1885 Art Loan Exhibition, New York
1889 Second Competitive Fan Exhibition, Draper's Hall, London
1890 Fan exhibition, Worshipful Company of Fan Makers, Draper's Hall, London
1891 Alte und Neue Fächer, Badischer Kunstgewerbe-Verein, Karlsruhe
1891 Fan exhibition, Budapest
1900 Eventails anciens faisant partie de la collection Lucien Duchet, Musée centennal et retrospectif, Exposition Universelle de 1900 Paris
1901 Fan exhibition, Stockholm
1902 Fan exhibition, Società delle Belle Arti, Florence
1903 Alte Fächer und Uhren, Palais des Ungarischen Ministeriums, Vienna
1903 Exposition de l'Ivoire, Musée Galliera, Paris
1905 Fächerausstellung, Friedman & Weber, Berlin
1907 Esposizione dell'ornamento feminile, l'évolution de l'art dans l'eventail, Palazzo Rospigliosi, Rome
1911 Fächerausstellung, Hohenzollernkustgewerbehaus, Berlin
1920 El abanico en Espana, Sociedad espanola de Amigos del Arte, Madrid
1926 Fans from the collection of Mrs Susanne Baldwin, formerly Mrs Lyman G. Bournisie, Art Institute of Milwaukee
1941 Another exhibition of the same – there was also an exhibition at Oshkosh at about this time
1949 Oude kant en waaiers, Museum Boymans-van Beuningen, Rotterdam
1954 The Fan Guild of Boston Exhibition, Wenham Museum, Boston, Massachusetts
1957 Viften – the fan, Copenhagen
1958 Eventails d'Espagne et de France, Palais Miramar, Cannes
1963 Exhibition of fans from the Leonard Messel Collection, Victoria and Albert Museum, London
1966 El abanico, Museo de Artes decorativas, Havana
1967 De waaier collectie Felix Tal, Centraal Museum, Utrecht
1971 Chinese fan paintings from the collection of Mr Chan Yee-pong, the University of Kansas Museum of Art, Kansas
1973 Eventails, Actualités – vie Parisienne, Musée Carnavalet, Paris
1974 Zauber des Fächers, Altonaer Museum, Hamburg
1974 An exhibition of the art of Chinese fan painting, Milne Henderson Gallery, London
1975 An exhibition of Nanga fan painting, Milne Henderson Gallery, London
1975 Special exhibition of paintings: folding fans from the collection 'The precious box of mists and clouds', National Palace Museum, Taipei
1975 Fans in Fashion, Temple Newsam House, Leeds, and Platt Hall, Manchester
1975 Fan exhibition, Naradowe Muzeum, Cracow
1976 The World of the Fan, The Fan Circle, Harris Museum and Art Gallery, Preston, Lancashire
1976 Nihon-ga Senmen-ga ten, an exhibition of painted fan leaves, Hokushin Garo, Tokyo

1978–9 Fans from the East, The Fan Circle, The Museum and Art Gallery, Birmingham, and The Victoria and Albert Museum, London
1979 Waaiers and Mode 18e eeuw tot Heden, Netherlands Kostuummuseum, The Hague
1979–80 Zauber des chinesischen Fächers, Museum Rietberg, Zurich, and Museum für Ostasiatische Kunst, Cologne
1980 Die Fächersammlung des Salzburger Museums, Carolino Augusteum, Salzburg
1980 Fans, prior to sale at Christie's, Foyle's, London
1980–1 Ostasiatische Fächerbilder und Fächer, Kunsthandel Klefisch, Cologne
1981 Fächer des 18. Jahrhunderts aus der Fächersammlung Gunnar A. Kaldewey, Düsseldorf – New York, San Francisco; The International Antiquarian Book Fair
1981 Fans in Fashion, Fine Arts Museum, San Francisco
1982 Fantastic Fans, Sharpstum Museum, Calistoga, California
1982–3 Fans and the Grand Tour, The Fan Circle, The Royal Pavilion Art Gallery and Museum, Brighton
1983 Fans – unfolding history, Richmond Museum, Richmond, Virginia
1983 Paris fans of the Belle Epoque, The Lachelin Archive, The Fan Circle, The Royal West of England Academy, Bristol
1983 Montgolfières et Carlines, Duvelleroy, Paris
1983–4 Fächer aus vier Jahrhunderten, Städische Sammlungen, Theodor-Haering-Haus, Tübingen
1983–4 Eventails de collections, Musée des Beaux Arts, Angers
1983–4 Fans, Vancouver Museum, Vancouver
1983–4 Indispensables Accessoires XVIe–XXe siècles, Musée de la Mode et du Costume de la Ville de Paris
1984 Fanfare, Renwick Gallery, National Museum of American Art, Smithsonian Institution, Washington D.C.
1984 Kompositionen im Halbrund, Fächerblätter aus vier Jahrhunderten, Staatsgalerie Stuttgart, Graphische Sammlung, Stuttgart
1984 The indispensable fan, the story of the fan in society, City Art Centre, Edinburgh
1985 L'éventail, miroir de la Belle Epoque, Palais Galliera, Paris
1985–6 Royal fans, The Fan Circle International, South Wiltshire Museum, Salisbury; Guildford; Harewood House, Leeds (2 catalogues)
1985–6 Ivory, Feather and Lace, Museum of London
1986 Quand l'éventail devient prestigieux, Louvre des Antiquaires, Paris
1986 Fächer der Welt, Welt der Fächer, Museum der Stadt Weinheim, Weinheim
1986 Fan, Tolsey Museum, Burford
1986–7 Ein Hauch von Luxus, Fächer und Fächerentwürfe aus vier Jahrhunderten, Heimatmuseum, Lippstadt
1987 Folding fans in the collection of the Cooper-Hewitt Museum, New York
1987 Fächer, Kunst und Mode aus fünf Jahrhunderten, Bayerisches Nationalmuseum, Munich
1987–8 Eventail, Musée d'art et d'histoire, Geneva
1988 Unfolding beauty – the art of the fan, Museum of Fine Arts, Boston
1988 Fans à la Mode, Orlando Museum of Art, Texas
1989 Modes & Revolution, Musée de la Mode et du Costume de la Ville, Paris
1989 L'éventail à tous vents, Louvre des Antiquaires, Paris
1989 Un Soffio di vanità, Ventagli dal XVII al XX secolo, Gran Caffè Pedrocchi, Padua
1989–90 Der Schönen Blicke Zügel, Der Europäische Faltfacher – Verden und Vandel, Schlöss Bruchsal, Karlsruhe
1990 Fans, Milton Manor, Oxon

MUSEUMS DISPLAYING FANS

AUSTRALIA
Melbourne: National Gallery of Victoria
Sydney: Power House
AUSTRIA
Salzburg: Österreichisches Museum für Angewandte Kunst
Vienna: Museum für Völkerkunde Hofburg (16th-century
    Ambros feather Mexican ceremonial fan – part of the
    Montezuma sent by Cortez to Charles V in 1524)
BELGIUM
Brussels: Musées Royaux d'Art et d'Histoire
FRANCE
Arles: Museé de l'Arletan
Angers: Château du Plessis Bourru (property of the Dowager
    Duchess of Dalmatia)
Bordeaux: Musée des Arts Decoratifs
Lyons: Musée des Tissues; Musée Lyonnais des Arts Décoratifs
Paris: Carnavalet; Musée des Arts Décoratifs (very fine
    collection); Musée de Cluny; Musée de la Mode et du Costume
St-German-en-Laye: Musée du Prieuré
GERMANY
Berlin: Konigl Museum für Völkerkunde (Chinese and other
    fans from the Karlsruhe exhibition
Frankfurt: Museum für Kunsthandwerk
Hamburg: Atoner Museum, Museum für Kunst und Gewerbe
    (Kokoshka fans)
Munich: Bayerischen Nationalmuseum
Nuremberg: Germanisches Nationalmuseum
GREAT BRITAIN
Aylesbury: Waddesdon Manor (small but select collection of 18th-
    and 19th-century fans); Buckinghamshire County Museum
Bath: Assembly Rooms
Bedford: Cecil Higgins Art Gallery (mainly 19th and 20th
    centuries)
Birmingham: City Museum and Art Gallery
Bournemouth Museum
Bradford: Bolling Hall Museum
Brighton: Art Gallery and Museum
Bristol: Blaise Castle Folk Museum
Cambridge: Fitzwilliam Museum (Rosse Collection)
Cardiff: Amgueddfa Werin Cymru (Welsh Folk Museum);
    St Fagin's Castle
Chertsey Museum
Chester: Grosvenor Museum
Christchurch: Red House Museum and Art Gallery (100 fans)
Coventry: Herbert Art Gallery and Museum
Dunfermline: Carnegie Dunfermline Trust, Abbey Park House;
    Pittencrieff House Museum
Durham: Gulbenkian Museum (Japanese)
Edinburgh: Royal Scottish Museum; National Museum of
    Antiquities (fan belonging to Mary Queen of Scots)
Exeter: Royal Albert Memorial
Glasgow: Camphill Museum; The Art Gallery and Museum
    (Burrell Collection)
Halifax: Bankfield Museum
Harrogate: Royal Pump Museum
Hartlebury Castle: Worcestershire County Museum
Hereford: City Library and Museum; Churchill Gardens
    Museum
Hitchin: Museum and Art Gallery
Hull: The Georgian House
Leeds: Lotherton Hall

Leicester: Wygston's House
Letchworth: Museum and Art Gallery
Lincoln Museum
London: British Museum, Dept of Prints and Drawings,
    Schreiber Collection (particularly for printed fans); Horniman
    Museum (ethnic); Museum of London; Museum of Mankind;
    Vestry House Museum of Local History and Antiquities;
    Victoria and Albert Museum
Luton: Museum and Art Gallery
Maidstone: Museum and Art Gallery
Manchester: Gallery of English Costume, Platt Hall
Newcastle-upon-Tyne: Laing Art Gallery and Museum
Norwich: Stranger's Hall
Nottingham: City Museum and Art Gallery
Oxford: Pitt-Rivers Museum; Ashmolean (oriental)
Peterborough Museum and Art Gallery
Preston: Harris Museum
Reading Museum and Art Gallery
Salisbury and South Wiltshire Museum
Snowshill Manor (National Trust)
Warwick: St John's House
Weybridge Museum
Wisbech Museum of the Fens
Worthing Museum
York Castle Museum
HOLLAND
Amsterdam: Rikjsmuseum
Rotterdam: Museum Boyman van Beuringen
    (Monchy Collection)
The Hague: Netherlands Kostuummuseum
ITALY
Florence: Museo Stibbert
Milan: Castello Storcesco, Museo Poldi Pozzoli
Naples: National Museum
Rome: Museo di Arte Orientale e Museo di Palazzo Venezia
Venice: Correr Museum
PORTUGAL
Lisbon: Museo Nacional dos Coches
SPAIN
Barcelona: Museo Cambo e Museo de Indumentere (Colecion
    Rocamora)
Madrid: Fernan-Nunez Palace (fine 18th-century collection);
    Aranjuez
SWEDEN
Stockholm: National Museum
SWITZERLAND
Geneva: Musee d'Art et d'Histoire
Noyon: Château de Basel, Kirschgaten Museum
Winterthur: Regional Museum Lindengut
Zurich: Swiss National Museum
UNITED STATES OF AMERICA
Boston: Museum of Art (Esther Oldham Collection)
New York: Metropolitan Museum of Art (Mrs William Randolph
    Hearst Collection); Brooklyn Museum; Cooper-Hewitt
    Museum
Orlando Museum of Art
Philadelphia Museum of Art
Phoenix Art Museum
Richmond: Valentine Museum
U.S.S.R.
Leningrad: Hermitage Museum

BIBLIOGRAPHY

Alexander, Helene, *Fans* (1984).

Alexandre, *Catalogue des éventails peints par les principaux artistes modernes* (Hotel Drouot, Paris, May 1875).

Allemagne, H.R.d', *Les Accessories du Costume et du Mobilier* (Paris, 1928).

Archaelogical Institute of America, Monographs on Archeology and Fine Arts I: Eitner, I.E.A., *The Flabellum of Tournus*, (1944).

Armstrong, Nancy, *A Collector's History of Fans* (1974).

Armstrong, Nancy, *The Book of Fans* (Colour Library International, 1979).

Art Journal, *Catalogue of the 1851 Exhibition*, p.313.

Art Journal (1875), p.103.

Arundel Society, *Fans of All Countries* (1871).

Bapst, Germain, *Deux Eventails du Musée du Louvre* (Paris, (1882).

Bars, Carlos M. (with Juan Escoda), *Eventails Anciens* (Lausanne, 1957).

Bennet, A.G. and Berson, R., *Fans in Fashion*, exhibition catalogue (Fine Arts Museum, San Francisco, 1981).

Bennet, A.G., and Berson, R., *Unfolding Beauty – The Art of the Fan*, exhibition catalogue (Museum of Fine Arts, Boston, 1988).

Bijutou Senshu, *Selected Objects of Japanese Art*, VI (1973): *Folding Screens with Fan Patterns in the Nanzeu-ji, Kyoto, Kano School*.

Blondel, Spire, *Histoire des Eventails et les notices sur l'ecaille, la nacre et l'ivoire chez tous les peuples et à toutes les Epoques* (Paris, 1875).

Boehn, Max von, *Das Beiwerk der Mode* (Munich, 1928).

Bojani, Count F. de, *Eventails Anciens* (sale, Brussels, 20 December 1912; ill.).

Bordez, M.F. *Fabrication des Montures d'Eventails à Ste. Geneviève* (1875).

Bouchot, Henri, 'L'Histoire par les Eventails Populaires', *Les Lettres et Les Arts* (Paris, January and July 1883).

Brno, Moravska Galerie, *Sperky, vjire* (Jewels, fans and miniatures in the collection of the gallery; Brno, 1968).

Burges, W., 'Notices of the precious objects presented by Queen Theodolinda to the church of St John the Baptist, at Monza', *Archaeological Journal* XIV (1857), 8.

Buss, George, *Der Fächer* (Düsseldorf, 1904).

Catalani, Carla, *Waaiers* (Bussum: Van Dishoeck, Holland, 1973).

Cherpentier et Fasquelle, *Un Siècle de Modes Féminines, 1794–1894* (Paris, 1895).

Chiba, Reiko, *Painted Fans of Japan; 15 Noh Drama Masterpieces* (Rutland, Vt, and Tokyo, 1962).

Chü Ch'ao tso-p'in hsüan-chi (A Selection of Fan Paintings by Chü Ch'ao; Canton, 1962).

Ch'u, Family, *T'ieh-chin T'ung-chien-lou ts'ang sang chi chin (Collection of Decorative Fan Mounts in the T.T. Collection in the Ch'ü family library of Chiang-Shu*; Shanghai, 1937).

Collection Lucien Duchet, *Catalogue des éventails anciens faisant partie de la classe 86 (group 13)* (Exposition Universelle, Paris 1900).

Collins, Bernard Ross, *A Short Account of the Worshipful Company of Fanmakers* (1950).

Commoner, Lucy A., *Folding Fans in the Collection of the Cooper-Hewitt Museum*, exhibition catalogue (New York, 1987).

The Connoisseur: I, pp.92, 175 (on repair/restoration); 15, p.84 (illus. of Marquis of Bristol's coll.); 19, p.197 (illus.); 24, p.213 (Mrs Beauclerk's coll. – some fans came from Saxe-Weimar and Goldschmidt colls).

Coomaraswamy, A.K., *Arts and Crafts of India and Ceylon* (1913).

Cosway, M., 'English Fans' in *The Concise Encyclopaedia of Antiques*, IV (1959), 298.

Croft-Murray, Edward, 'Watteau's Design for a Fan-Leaf', *Apollo* (March 1974).

Crossman, Carl L., *The China Trade* (Princeton, 1972), CH. II,

Crystal Palace, Tallis' History and Description of the, I (1851), pp.214–19.

Cust, Lionel, *Catalogue of the Collection of Fans and Fan Leaves, presented to the Trustees of the British Museum by Lady Charlotte Schreiber* (1893).

Dawes, Leonard, 'The Nicely Calculated Flutter of the Fans', *Antiques Dealer and Collector's Guide* (March, 1974).

Delpierre, Madeleine(ed), Falluel, Fabienne, Maignan, Michel, and Trugan, Rosie, *l'Eventail Miroir de la Belle Epoque*, exhibition catalogue (Palais Galliera, Paris, 1985).

Diderot, D. and d'Alembert, 'Eventaille', *Encyclopédie* (1765).

Dubose, Jean Pierre, *Peintres Chinois du XVIe Siècle: Wen Tchang ming et son école: presentation et étude de quelques aeuvres (éventails et feuilles d'album)* (exhibition at Galerie Maurice, Lausanne, 1961).

Dunn, D., 'On Fans', *Connoisseur* (1902).

Durian-Ress, Saskia, and Heller-Winter, Elizabeth, *Fächer, Kunst und Mode aus Fünf Jahrhunderten*, exhibition catalogue (Bayerischen Nationalmuseum, Munich, 1987).

Eeghen, I.H. van, 'De Waaier en de Poolse successieoorlog', *Tijdschrift voor Geschiedenis* (Amsterdam, 1963).

Eeghen, I.H. van, 'De Amsterdamse Waaierindustrie, ae XVIIIe eeuw', *Amstelodamum* (1953).

Eeghen, I.H. van, 'De Watergraafsmeer op een waaier van 200 jaar geleden', *Amstelodamum* (1958).

Erler, M., 'Der Moderne Fächer', *Kunstgewerbeblatt* (September 1904).

Falluel, F., and The Fan Circle International, *Paris Fans of the Belle Epoque*, exhibition catalogue (Bristol, 1983).

Fan Circle, *Bulletin* (1975 onwards).

Fan Circle in association with the Victoria and Albert Museum, *Fans from the East* (1978).

Fan Guild of Boston, *Fan Leaves* (Boston, 1961).

Flory, M.A., *A Book about Fans* (1895).

Fraipont, G., *l'Árt de composer et de peintre l'éventail, l'ecran, le paravent* (H. Laurent, Paris, 1896).

Gibson, Eugenie, 'The Golden Age of the Fan', *The Connoisseur*, 56 (1920), Gibson, Eugenie, 'Queen Mary's Collection', *The Connoisseur*, 78 (1927).

Giles, H.A. 'Chinese Fans', *Fraser's Magazine* (May 1879); reprinted in Giles, H.A. *Historic China and other Sketches* (1882).

Gostelow, Mary, *A Collector's Guide to Fans* (1976).

Grand Cateret, John, *Vieux Papiers, vieilles images: cartons d'un Collectionneur.* (Paris, 1896).

Great Exhibition, Official Catalogue of the (1851).

Green, Bertha de Vere. *A Collector's Guide to Fans over the Ages* (1975).

Gros, Gabriella, 'The Art of the Fan Maker', *Apollo* (January 1957).

Gubanov, G., Vasilevskaja, N., and Malachou, S., *Western European fans from the art collections of the Soviet museums* (Leningrad, 1972).

Hallwyska Samlingen, *Boskrifuande fürteckning XXVI–XXX/ Solfjadrar Minnen of personlige upplefrelser, Redem innen Minner of almänne handelser minnen frankrinsären 1914–1919* (Stockholm, 1937).

Hammar, Britta (Swedish), 'Fans of the 18th. Century', *Kulturen*, 1976; trans. Marion Maule, *Fan Circle Newsletter* (9 October 1978).

Harari, Ralph A., *The Harari Collection of Japanese Paintings and Drawings* (J. Hillier, 1973).

Hay, John, 'Chinese Fan Painting and the Decorative Style', *Colloquies on Art and Archaeology in Asia*, No. 5 (Percival David Foundation, 1975).

Hay, John, *An Exhibition of the Art of Chinese Fan Painting* (Milne Henderson Gallery, 1974).

Heath, Richard, 'Politics in Dress', *The Woman's World* (June 1880).

Hiroshi, Mizuo, *Edo Painting* (New York and Tokyo, 1972).

Hirshorn, A.S., 'Mourning Fans', *Antiques*, CIII (1973), p.801.

Holme, C. 'Modern Design in Jewellery and Fans', *Studio* (1902).

Holt, T.H., 'On Fans, their use and antiquity', *Journal of the British Archaeological Association*, XXVI (1870).

Honig, G.J., 'Wat een antieke waaier ons ken vertellen met betrekking tot de Nederlandse walvisvaart', *Historia* (1942).

Hughes, Therle, 'Lady Windermere's Fan', *Country Life* (20 December 1973).

Hughes, Therle, 'Storm Dragons and Plum Blossom', *Country Life* (15 June 1972).

Irons, Neville John, *Fans of Imperial China* (Hong Kong, 1982).

Irons, Neville, John, *Fans of Imperial Japan* (Hong Kong, 1982).

Jackson, Mrs. F. Nevill, 'The Montgolfiers', *The Connoisseur*, 25 (1909).

Jong, M.C. de, *Waniers & Mode 18e eeuw tot Heden*, exhibition catalogue (Netherlands Kostuummuseum, 1979).

Kammerl, Christl, *Der Fächer Kunstobjekt und Billetdoux* (Munich, 1989).

Kansas, University of, *Chinese Fan Paintings from the Collection of Mr. Chan Yee-Pong* (Lawrence, Kansas, 1971).

Kendall, B., 'Concerning Fans' *The Connoisseur*, 7 (1903), p.4 (1902 loan exhibition, Società delle Belle Arti, Florence, from colls of Queen Margherita, Duke and Duchess of Aosta, and Dowager Duchess of Genoa).

Kimura, Sutezō, *Edo Katuki uchiwa-e: Genroken-Enkyō ren (Kabuki fan prints from Edoi Genroku to Enkyō periods 1688–1748;* commentary by S.K. and S.I. Miyas, Tokyo, 1962).

Kiyoe, Nakamura, *Ogi to Ogie (Fans and Fan Painting;* Kyoto, 1969).

Ku Kung Po Wu-Yuan-ku kung ming shan chi (Collection of famous fans from the Imperial Palace, I–IX ed. I Pei-chi, x ed. Shau Chang; Peking, 1932–35).

Kyoto Senshoku, *Bunka Kenkyukai – X korin-ha semmen gashū (Collection of Fan Painting of the Kōrin School),* ed. by the Society for Research into Dyeing and Weaving (Kyoto, 1967).

Leary, E., *Fans in Fashion,* exhibition catalogue (Temple Newsam, Leeds and Platt Hall, Manchester, 1975).

Leningrad: Hermitage, *Zapadnoeuropaiskie veera XVIII–XIX vv, iz sobraniya . . . Ermitzha: Katalog vremennoi vai stavhi (Western European Fans of the 18th and 19th centuries in the Hermitage Collection),* catalogue of a temporary display with introductory essay by M.I. Torneus (1970).

Linas, Ch. de, 'Les disques Crucifères, le flabellum et l'umbrella', *Revue de l'art Chrétien* (1883).

Marcel, Gabriel, 'Un Eventail Historique du Dix-huitième Siècle', *Revue Hispanique* VIII (1901).

Margary, Ivan D., 'Militia Camps in Sussex 1793; and a Lady's Fan', *Sussex Archaeological Records* CVII (1969).

Mayer, Carol, E., *Fans,* exhibition catalogue (Vancouver Museum, Vancouver, 1983).

Mongot, Vincente Almela, *Los Abanicos; Fans of Valencia* (Spain, n.d.).

Mourey, Gabriel, Vallance, Aymer, *et al., Art Nouveau Jewellery and Fans* (New York, 1973).

National Encyclopaedia, The, V (c.1880).

Muller Krumbach, Renate, *Alte Fächer* (Weimar, 1988).

Niven, T., *The Fan in Art* (New York, 1911).

North, Audrey, *Australia's Fan Heritage* (1985).

Ohm, Annaliese, art 'Fächer', in *Reallexikon zur Deutschen Kunstgeschichte* (Stuttgart, 1972).

Palliser, Mrs. Bury, *History of Lace* (1865).

Parr, Louisa, 'The Fan', *Harper's Magazine* (August 1889), pp. 399–409.

Percival, MacIver, *The Fan Book* (1920).

Percival, MacIver, 'Some Old English Printed Paper Fans', *The Connoisseur*, 44, p. 141.

Perthuis, Francoise de, and Meylan, Vincent, *Eventails* (Paris, 1989).

Petit, Edouard, *Le Passé, le présent et l'avenir: Études, souvenirs et considérations sur la fabrication de l'éventail* (Versailles, 1895).

Powell, B.H. Baden, *Handbook of the Manufacturers and Arts of the Punjab* (1872).

Redgrave, S., Preface to South Kensington *Catalogue of the Loan Exhibition of Fans* (1870).

Reig y Flores, Juan, *La Industria Abaniquera en Valencia* (Madrid, 1933).

Rhead, G.W. Woolliscroft, *The History of the Fan* (1910).

Rhead, G.W. Woolliscroft, *The Connoisseur*, 55 (1919) and 58 (1920) (on Messel coll.)

Robinson, F. Mabel, 'Fans', *The Woman's World* (January 1889).

Rocamora, Manuel, *Abanicos Historicos y Anedóticos* (Barcelona, 1956).

Rondot, Natalis, *Rapport sur les Objets de Parure, de Fantaisie et du Goût, fait à la Commission Française du Jury International de l'Exposition Universelle de Londres* (Paris, 1854).

Rosenberg, Marc, *Alte und Neue Fächer aus der Wettbewerbung und Ausstellung zu Karlsruhe* (Vienna, 1891).

Rusconi, John, *The Connoisseur*, 18, p.96 (on coll. of Queen Margherita of Italy).

Sacchetto, Amelia Filippini, and Testoni Sassi, Lidia, *Un Soltio di Vanità,* exhibition catalogue (Padua, 1989).

Salwey, Charlotte M., *Fans of Japan* (1894).

Schreiber, Lady Charlotte, *Fans and Fan Leaves – English* (1888).

Schreiber, Lady Charlotte, *Fans and Fan Leaves – Foreign* (1890).

Seligman, G.S. and Hughes, T., *Domestic Needlework* (1938).

Sociedad Española de Amigos del Arte, *Exposicion de 'El Abanico en España' 1920* (Madrid, May–June) (by D. Joaquin Ezqueria del Beyo).

Spielmann, H., *Oskar Kokoschka: Die Fächer für Alma Mahler* (Hamburg, 1969).

Standen, E.A. 'Instruments for Agitating the Air', *Metropolitan Museum of Art Bulletin*, XXIII (March 1965).

Strange, Edward F., *The Colour Prints of Hiroshige* (Cassell, 1925).

'Studio-Talk: Fans', *Studio*, XXXVII (1906), No. 158; XLV (1909), No. 190.

Suoboda, Christa, *Die Fächerausstellung des Salzburger Museums Caroline Augusteum,* exhibition catalogue (Salzburg, 1980).

Taipei, Taiwan, *Masterpieces of Chinese Album Painting in the National Palace Museum* (1971).

Tal, Felix, *De Waaier Collectie Felix Tal* (exhibition; Utrecht, 1967).

Thornton, Peter, 'Fans', *Antiques International* (1966).

Thornton, Peter, 'Une des plus belles collections du monde' (Leonard Messel Collection), *La Connoissance des Arts* (April 1963).

Udstillingen ny kongensgade i til Fordel for Selskabet til Haanderbejdets Fremme, Viften, *The Fan* (Copenhagen, 1971).

Uzanne, Octave, *The Fan* (Eng. trans. 1884).

Vecellio, *Habiti Antichi et Moderni,* etc. (1590).

von Keudell, Baroness, *The Connoisseur*, 13 (1905), p.152 (notes on coll. of Miss Moss, Fernhill, Blackwater).

Volet, Maryse, *l'Imagination au Service de l'Eventail* (Vezenaz, 1986).

Volet, Maryse, and Beentjes, Annette, *Eventails Collection du Musée d'Histoire de Genève,* exhibition catalogue (Geneva, 1987).

Walker, Robert, *Catalogue of The Cabinet of Old Fans* (Sotheby, Wilkinson & Hodge, 1882).

Wardle, Patricia, 'Two Late Nineteenth Century Lace Fans', *Embroidery* XXI (summer 1970), no. 2.

White, Palmer, *Poiret* (1973).

Worshipful Company of Fanmakers: Report of Committee, etc., *Competitive Exhibition of Fans held at Draper's Hall* (July 1878).

'X, Mme.', sale catalogue, *Objets d'Art, éventails, Louis XV dentelles,* etc., Hôtel Drouet (Paris, 1897).

Zauber des Fächer aus den Besitz des Museum, Altonaer Museum, Hamburg (1974).

INDEX

Page numbers in *italic* refer to the illustrations